SUSTAINABLE ENERGY SYSTEMS

ERRATUM

Dovers *Sustainable Energy Systems*
0 521 43099 2 Hardback
0 521 47757 3 Paperback

p. 33, key to figure 2.3

Sectors

1 Agriculture, forestry, fishing, hunting	15 Other machinery and equipment
2 Mining	16 Miscellaneous manufacturing
3 Meat and milk products	17 Electricity
4 Other food products	18 Gas
5 Beverages and tobacco	19 Water
6 Textiles, clothing, footwear	20 Construction
7 Wood, wood products, furniture	21 Wholesale and retail, repairs
8 Paper products, printing, publishing	22 Transport, storage, communication
9 Chemicals	23 Finance, property, business services
10 Petroleum and coal products	24 Residential property
11 Non-metallic mineral products	25 Public administration, defence
12 Basic metals, products	26 Community services
13 Fabricated metal products	27 Recreational, personal services
14 Transport equipment	

FUNDAMENTAL QUESTIONS PROGRAM

This book is published as an outcome of the Fundamental Questions Program of the Centre for Resource and Environmental Studies, The Australian National University. The Program began in 1988, focussing interdisciplinary research efforts on the problem of achieving ecological sustainability. Energy was one of the original theme areas.

The Program promoted research and systematic discussion on the implications of this fact for the future of human society. The design of the Program reflected appreciation of the fact that the biosphere, as a system capable of supporting humankind, will not tolerate indefinitely the pattern of resource and energy use characteristic of present-day society. An aim of the Program was to present the research outcomes to a wider audience. This book seeks to fulfil this aim in the area of sustainable energy.

Information about the Fundamental Questions Program and publications arising from it can be obtained from the Centre for Resource and Environmental Studies, Australian National University, ACT 0200.

SUSTAINABLE ENERGY SYSTEMS

Pathways for Australian Energy Reform

Edited by

STEPHEN DOVERS

Centre for Resource and Environmental Studies,
The Australian National University

CAMBRIDGE
UNIVERSITY PRESS

CAMBRIDGE UNIVERSITY PRESS
Cambridge, New York, Melbourne, Madrid, Cape Town,
Singapore, São Paulo, Delhi, Tokyo, Mexico City

Cambridge University Press
The Edinburgh Building, Cambridge CB2 8RU, UK

Published in the United States of America by Cambridge University Press, New York

www.cambridge.org
Information on this title: www.cambridge.org/9780521477574

© Cambridge University Press 1994

First published 1994
Re-issued 2011

A catalogue record for this publication is available from the British Library

Library of Congress Cataloguing in Publication Data

Sustainable energy systems: pathways for Australian energy reform/
edited by Stephen Dovers.
Includes index.
1. Power resources - Australia. 2. Energy policy - Australia.
3. Renewable energy sources. I. Dovers, Stephen.
TJ163.25.A8S87 1994
333.79′094-dc20   94-14986
                                                        CIP

ISBN 978-0-521-43099-9 Hardback
ISBN 978-0-521-47757-4 Paperback

Contents

Part 4   Towards sustainable energy systems

Figures

Tables

Contributors

Michael Common is a Senior Fellow at the Centre for Resource and Environmental Studies, The Australian National University, Canberra; and Reader, Department of Environmental Economics and Environmental Management, University of York.

Mark Diesendorf is National Campaign Convener, Energy and Transport, Australian Conservation Foundation, based in Canberra.

Stephen Dovers is a Research Officer at the Centre for Resource and Environmental Studies, The Australian National University, Canberra.

Ian Lowe is Head of the School of Science, Griffith University, Brisbane.

David Mills is a Senior Research Fellow at the Department of Applied Physics, University of Sydney.

Peter Newman is Director of the Institute for Science and Technology Policy, Murdoch University, Perth.

Hugh Saddler is a Director of Economic and Energy Analysis Pty Ltd, Canberra.

Units and conversions

In this volume, the units used are joules (J) and Watts (W), or multiples thereof. These units are defined in Chapter 1. The multiples used are as follows:

kilo (k)	$\times 10^3$	kJ, kW, kWh
mega (M)	$\times 10^6$	MJ, MW, MWh
giga (G)	$\times 10^9$	GJ
peta (P)	$\times 10^{15}$	PJ

The following allows for conversion between commonly used energy units:

1 calorie = 4.2 joules
1 million tonnes oil equivalent (MTOE) = 41.9 petajoules (PJ)
1 million tonnes coal equivalent (MTCE) = 28.8 petajoules (PJ)
1 British Thermal Unit (BTU) = 1.055 kilojoules (kJ)
1 kilowatt-hour (kWh) = 3.6 megajoules (MJ)

Preface

This book addresses the very urgent imperative of creating energy systems that are both ecologically sustainable and humanly desirable. The intention has been to integrate a number of views of energy: the historical context; a systemic view of energy across society; longer-term visions of what might be possible; pragmatic analyses of what can be achieved in a practical sense; and considerations of appropriate policies. The emphasis is both on surveying available technologies, and on identification of the non-technical barriers which influence their adoption or avoidance. Another aim has been to communicate the material in a manner accessible to a wider readership. The primary focus here is on Australia, but much of what is discussed can be applied more widely. While the coverage is not complete, the pathways of energy reform described here offer substantial opportunities if society is prepared to follow them.

The book is unapologetically reformist. Not too much space is devoted to discussing the basis of the problem – the great human and ecological dangers of continuing with our present energy systems are well enough established. After some fundamentals are covered in the first two chapters, the book moves on to what we can do. Part 2 deals with conserving energy and using it more efficiently. Part 3 describes renewable energy options. Part 4 considers the policy measures available to us.

Some acknowledgements are required. To a large degree, I became convener of the energy theme in the Fundamental Questions Program and thus editor of this book, by default and with little grounding in, but none the less a broad enthusiasm for, the complex area of energy. The help and support of other people has been needed and appreciated. The research program was enabled only through the direction and inspiration of Stephen Boyden, whose comments on the manuscript were greatly appreciated. The support of the Director of CRES, Henry Nix, was crucial. The contributors to this volume provide it with its substance, and all have done this in the face of many other pressing responsibilities. The figures were drawn by Kevin Cowan of The Australian National University.

S. DOVERS

PART ONE

Introduction and background

CHAPTER 1

Introduction: the issue of energy

STEPHEN DOVERS

An energetic world

Energy is the basic currency of the biosphere. Without energy, *change* is impossible. Without energy, matter cannot be moved or transformed; quite simply, nothing can be done. Picture an animal or group of animals chasing prey in a natural ecosystem. Should these animals expend more energy than they eventually gain through eating the food they find, they will surely die.

For the vast bulk of the history of life on Earth, over many hundreds of millions of years, this was an unbending law for all living things. An organism survived on that fraction of the energy flowing through its environment that it could capture and use within its own body. In the relatively very recent past, however, one species of animal, *Homo sapiens*, has managed to escape the constraints of this law. When humans first began using fire, they developed an attribute that was to change the course of planetary history – the ability to tap sources of energy external to their own bodies. And so our species started being different from other animals, doing unique things like creating heat, cooking food, and changing the landscape through burning.

Before the use of fire, our ancestors had available to them only that energy that flowed through their own bodies, taken in as food and expended through metabolic processes just like any other animal (this can be called *somatic energy*). An average figure for somatic energy for an adult human being is about 10 megajoules (MJ) per day or 3.65 gigajoules (GJ) per year (this figure varies with body weight, activity level, climate, and so on). The use of fire in hunter–gatherer societies provided an additional source of energy – *extrasomatic energy* – which

2

probably about doubled the energy budget of a human being to 7 or 8 GJ per year (Boyden 1987). The domestication of certain animal species and their subsequent use as a source of draught power was a later and highly significant development.

When we consider the history of technological development, we tend to think of *devices* such as wheels, internal combustion engines and telecommunication equipment. But the underlying basis of technological and economic development has been the introduction of new sources of usable energy. It has been said that '... from one perspective, history is the story of the control over energy sources for the benefit of society' (Goldemberg *et al.* 1988: 2). Ponting (1992) supports this view, describing the use of fossil fuels as the second great transition in human history, after the advent of farming. Since the discovery of fire, humans have added much more potent sources of energy to their repertoire – coal, oil, uranium, various gases, the power of moving wind and water. These have been the ways around the basic energy constraint of life. The increase in the use of energy by humans has quite literally been explosive. The global average use of energy is now about 70 GJ per year, and as high as 400 GJ in some countries.

This book is based on the view that the supply and use of energy by humankind is a crucial problem that must be confronted if we are to achieve an ecologically sustainable and humanly desirable future. Using energy has allowed us to do things that have both enabled and enriched the lives of countless millions of people – cooking food, growing more food, keeping warmer or cooler in harsh climates, building homes, communicating, making machines and other objects, providing transport and other services – all such things require energy. But this has not been without cost.

The issue of energy is of overwhelming importance for five basic reasons:

1. Because it is required to move or transform matter, energy reflects the *level of physical activity* (i.e. the scale of interaction with the surrounding environment) of any defined entity, be it an individual, a firm, a region, a society, or the global human population.
2. Because of this, energy is a necessary and irreplaceable input of *all* sectors in a modern economy; the flow of energy through a production system may be reduced relative to unit output, but never eliminated. Energy is one of the few categories of resources that are common to all human activities (the only others arguably being land and labour) – *everything* consumed or used by humans has an energy component.
3. The waste products of the fossil-fuel-based energy systems of

industrialised countries are the primary cause of predicted global environmental change resulting from the enhanced greenhouse effect (particularly emissions of carbon dioxide).

4. These same energy systems are also the source of the bulk of the troublesome air pollutants in industrial societies (oxides of carbon, sulfur and nitrogen).

5. The bulk of energy used is from non-renewable sources such as coal, oil, natural gas or uranium. Very little commercial energy is from renewable sources. The eventual exhaustion of these resource stocks is therefore inevitable and a cause for concern.

These points need to be kept in mind lest we slip into treating energy as just another commodity. Energy is a pervasive challenge in questions of sustainability (Peet 1992 offers an excellent and detailed discussion).

In discussing a society's impact on the environment, the equation of Wasi (1991) captures the essence of the problem:

$$E = NB$$

where E is environmental impact, N is the number of people, and B is their behaviour. *Behaviour* is complex, and the hardest variable to define, being the totality of actions of an individual or group of people. Behaviour is shaped by a variety of cultural, political, technological and economic factors. In terms of impact on the environment, behaviour can be related to the consumption of natural resources and the use of the assimilative capacity of the environment to absorb wastes. In terms of energy use, this is important as it recognises a fundamental point: a country small in population (like Australia) can have a disproportionately large impact, per person, on the global environment due to its very high per capita rate of consumption.

It can be argued that energy use, per capita or total, is a useful single indicator of the 'environmental load' of a person or a society (although, of course, it is not the only indicator). Since the time that people first started farming some twelve thousand years ago, the figures for total energy use would indicate that *the environmental load that humans place on the biosphere has increased more than ten-thousand-fold*. It is still rising in line with increases in both per capita energy use and population. The World Commission on Environment and Development (the 'Brundtland' commission) noted that, given existing technologies, bringing the whole world's resource and energy consumption up to that which the industrialised countries now enjoy would increase total energy consumption five-fold. The Commission was of the opinion that the biosphere could not tolerate the implications of this (WCED 1987).

So we face the challenge of whether human society can create systems of energy supply and use which are ecologically sustainable in the long term and can allow us to satisfy the requirements for the health and well-being of all people. Further energy-intensive development has become the traditional way of meeting our needs. Given that human society is currently *not* ecologically sustainable, and that the health and well-being needs of all people are most certainly not being met, the challenge is clearly one of great magnitude (see, for example, WCED 1987).

The various chapters in this book seek first to make clear why the issue of energy is so important, and then to point towards some directions that might be taken in accepting this challenge.

The five points above can provide an initial basis upon which to assess energy systems and energy futures. In terms of the first point, a reduction in the total human energy load would appear desirable. Point 2 focuses attention on increasing the effectiveness of using energy to produce a given commodity or service. Points 3–5 emphasise the need for energy systems which minimise the production of wastes and maximise the renewability of the energy resources relied upon.

To embark on a positive note, there is a fundamental truth that should be a touchstone in this search for direction. It may seem contrary to the importance placed here on energy, but in fact *no one really wants energy* (Patterson 1991). What we want, at the risk of speaking in platitudes, is health and happiness. We humans want the things that make life comfortable and rewarding, and energy is necessary for any of this – to eat, to keep sheltered, to travel to see friends and relatives, to recreate, and so on. Energy is but a means to these ends. The question is, can we achieve these ends with less of the means?

The nature of energy

Energy is a common word, used to mean many things in a multitude of contexts. We need to define early on the sense in which the word energy, and related terms, are used in this book (for more detail, see Odum and Odum 1976; Slesser 1982; Patterson 1991).

The word 'energy' comes from the Greek *ergon* (work) and *en* (in), and was proposed in the early nineteenth century by Thomas Young as, essentially, a term to describe the *ability to do work*. This, and related developments, represented a significant event in our ability to understand the physical world. The standard unit for measuring energy is the *joule* (J) or multiples thereof (other units are also used – see conversion table). All matter has an energy content, which can be quantified as its energy density at certain conditions and varies greatly (for example: oil, 42 MJ/kg; liquid hydrogen, 120 MJ/kg; straw, 18 MJ/kg). Energy takes many forms, including, for example, the *kinetic* energy of a body in motion, the *chemical* energy released in metabolising food, or the *potential* energy inherent in an object poised to fall or a chemical that may react.

Related concepts are work and power. *Work* is a measure of effort, and is also typically measured in joules. *Power* is the rate at which work is done and is measured in *watts* (W) or multiples thereof. A watt is equivalent to one joule per second. Thus power can be thought of as the *rate* at which energy is being converted. *Heat* is energy in transfer from one environment or system to another.

There are two basic laws that, as far as we are aware, represent absolute constraints that are crucial for understanding energy. Simply stated, these are:

- **The first law of thermodynamics** (alternatively the law of the conservation of energy), which states that energy can be converted into another form, or conserved for a length of time in one form – *but energy can never be created or destroyed.* Energy flowing into a system is either stored there or flows out. Thus the total amount of energy, in all its forms combined, remains constant.

    (This law forces us to accept that, although we always talk about energy 'consumption', energy is never actually consumed, but rather used. We simply transfer it from one form to another in processes that have, usually, some desired outcome.)

- **The second law of thermodynamics** (alternatively the law of the degradation of energy), which states that, in any process, some of the energy will be dissipated or degraded. This law is sometimes stated as meaning that any system tends towards increasing disorder, and is sometimes also called the *entropy* law.

    The best illustration of this law is the fact that no conversion of energy can be perfect – energy is 'lost' to the system, becoming lower quality energy that cannot be used by the system. Such a system might be an organism (about 90% of the energy in food is typically dissipated as waste heat); a car (only around a tenth of the energy in the fuel (petrol or gasoline) moves the car along, the rest is lost as waste heat, chemical energy in emissions, and friction); or a thermal electricity generation plant, where perhaps 10 J of energy in coal is combusted to produce 3 J of electricity. This 'loss' of energy to a system is tolerable, as long as the incoming energy supply is maintained, because some of the energy is converted into a more organised state, a higher quality, which the system desires. An organism 'wastes' energy as heat, but stores some as body tissue or uses some in metabolic processes, and thus maintains itself. A generation plant disperses most of the coal's energy content, but converts some into the incredibly useful and high quality form of electricity.

Critical to the issue of energy are its various sources. The biosphere is powered by solar energy from the Sun: it gains energy from the Sun and this is balanced by the eventual loss of energy back to space. A minute fraction of the Sun's energy reaches Earth and more than half of that is either reflected before reaching the surface, or is absorbed by the atmosphere. Of that reaching the Earth's surface, about 30% is reflected as heat, 21% drives the winds and about 40% is involved in the evaporation and condensation of water. Less than 1% is captured by plants and converted into chemical energy via *photosynthesis*, the process which underpins the existence of life as we know it.

Both natural ecosystems and human-managed ecosystems thus run on solar energy. Different ecosystems vary widely in the amount of energy that flows through them, and this is a measure of their biological productivity. Table 1.1 gives some examples of the extreme 'patchiness' of production in the natural world: its energy intensity. The variability of energy intensity in a human society is discussed in the next chapter.

Humans, like all other animals, 'eat' solar energy in the form of either plant or animal tissue, taking in both energy and nutrients. When humans first deliberately used fire, they were of course using solar energy, but it was energy that had been stored for a time in the tissue of those large plants that we call trees. The beauty of wood as a fuel was that it was portable – a source of energy that was mobile and controllable. The portability of energy is also central to modern human energy systems. Recent human history, especially from the industrial

Table 1.1 *Net plant production across selected ecosystems (megajoules per square metre)*

Ecosystem	Production
Extreme desert	0.06
Open ocean	2.4
Continental shelf	6.6
Lakes and streams	9.4
Temperate grasslands	9.4
Woodlands and shrublands	11.3
Agricultural land	12.2
Forest — coniferous	15.1
— temperate	24.5
— tropical	37.7
Estuaries, swamps and marshes	37.7
Average — for oceans	2.9
— for land	13.8
— for total Earth	6.0

Source: Adapted from Kormondy 1976.

revolution onwards, has been the story of utilising higher quality and more portable sources of energy.

Energy: recent history

Basic to the modern economy is a supply of *fuels*, which are those substances that we use as carriers of energy, and which allow us not only to produce vastly more of the *heat* that wood allowed us to enjoy, but also to convert large amounts of this into *mechanical* energy to perform tasks previously unthinkable.

For millennia, the main fuels used by humans were wood and other *biofuels* such as dung. This provided heat to warm themselves, to cook food, to bake clays or to smelt metals. They also discovered means of harnessing other sources of mechanical energy. At first, this was in the form of the somatic energy of draught animals and, later, the extra-somatic energy of moving air (sails, wind-driven mills) and moving water (water-driven mills). These sources of energy were very useful but not particularly reliable, of high quality, or portable.

Two forms of fuel exist: *chemical* (which can be burned to give off heat) and *nuclear* (which can undergo fission). Chemical fuels – coal, oil, gas, wood and other biofuels – are by far the more important in the global energy system, either used in their primary form or refined or converted into other forms (such as crude oil to petrol, or coal to electricity). All these chemical fuels are stored forms of solar energy – in wood, or that energy stored long ago in the dead bodies of organisms that we now call the *fossil fuels*: coal, oil and natural gas. Similarly, the smaller amounts of energy captured as hydro-electricity (derived from moving water), from wind or various forms of solar collectors are solar energy that we can convert. Nuclear energy is different, being the inner energy of atoms which humans now can release, and which has never before been a part of the solar-driven energy system of the Earth. Fuels may be *primary* or *derived*. For example, oil or coal may be combusted directly in a boiler as a primary fuel, or in a generation plant to produce electricity, a derived fuel.

An important difference between sources of energy is whether they are *renewable* or *non-renewable*. At present our society relies primarily on non-renewable forms of extrasomatic energy: oil, coal, gas and nuclear. Dependent on the rate of use and the size of the resource stock, such energy sources will at some stage be exhausted. The use of renewable forms of energy does not necessarily involve a run-down of a finite resource stock. This category includes sources such as solar, hydro, wind and tidal energy and (if managed properly) *biomass* energy sources such as grain alcohol or fuelwood. Apart from these stock aspects, renewable energy sources in general involve the production of far less troublesome waste products such as carbon dioxide or various air pollutants than do non-renewable sources. This is especially the case for solar, wind, hydro and tidal power, but somewhat less the case for biomass fuels. For both these reasons, renewable energy is preferable, from the standpoint of long-term supply and environmental and health issues.

The beginning of the modern phase of human history – *the high-energy phase* – can be dated as around the same as the beginning of what is more broadly known as the Industrial Revolution (Boyden 1987). The steam engine, at first powered by wood, was soon stoked with coal and gave humans significant amounts of portable, controllable mechanical energy for the first time. Later, the use of oil, the development of the internal combustion engine, and the discovery and use of electricity expanded the available supply of energy, making possible the heat transport machines and the communications systems which support modern civilizations.

In 1650 AD, prior to the Industrial Revolution, the somatic energy use of the world's human population was in the order of 2000 peta-joules (PJ) and the extrasomatic energy use about 4000 PJ. In 1990, total human energy use was about 360,000 PJ, about 95% of this being extrasomatic energy derived mostly from fossil chemical fuels.

Energy: problems, addictions and traps

As previously stated, the creation of new energy systems is a challenge of great magnitude. This is precisely because energy is so fundamental and important; because it is a basic ingredient of every action and process in a human society. The World Commission on Environment and Development (1987: 15) put it thus (*emphasis added*):

> A safe, environmentally sound, and economically viable energy pathway that will sustain human progress into the distant future is clearly imperative. It is also possible. *But it will require new dimensions of political will and institutional cooperation to achieve it.*

Particularly in the industrialised world, social and economic systems have evolved that are completely dependent upon unending and large inputs of fossil fuels. Many cities and towns are so structured that the lack of a motor car is a serious social and economic disadvantage. We are reliant on goods that are produced at some distance and so depend on energy-intensive transport systems. In fact, the spatial arrangemnts of industrial societies are very much a product of their energy systems

(Owens 1986). Industrial agriculture requires large energy inputs, our communications systems are based on electricity, and the manufactured products we consume, to varying intensities, require inputs of energy in their production. Modern societies have certainly become addicted to large quantities of cheap and convenient energy. Our social, economic, production and political systems are dependent on it. Moreover, the geopolitics of the modern world is strongly influenced by energy (Tsai 1989; Kapstein 1990).

But it has become increasingly obvious that, while it may have seemed a 'good idea at the time', this addiction has many drawbacks at first not obvious. Some of these were mentioned earlier, and are discussed in more detail in the next chapter. In this way, energy is a classic example of what has been dubbed a *social trap*, a term coined by Platt (1973). Social traps are defined by Brechner and Linder (1981: 29) as:

> ... situations where the short-term consequences for a behaviour are positive and the long-term consequences are negative.

Driving a car to purchase goods, cooking using electricity, heating a non-insulated home – these are convenient behaviours with positive short-term consequences. Air pollution, run-down of resource stocks, global climate change, urban congestion – these are the inconvenient long-term consequences. At a time before such conveniences were invented, humans obviously did without them, achieving their ends by some other means, but now most people in industrialised societies could scarcely envisage life in their absence. Once introduced, many innovations are adopted at an astoundingly rapid rate and become classic examples of the principle of *technoaddiction* (Boyden 1987).

The essence of a social trap is that it is indeed a trap; easy to enter but hard to escape. We have entered the trap and have benefited from enormous increases in energy use, but can now see the long-term consequences and are beginning to seek ways to avoid or lessen these. In social trap analysis, various solutions have been proposed (adapted from Brechner and Linder 1981) to:

- reduce the delay period between the behaviour or action and the negative consequence (that is, to make the consequence more immediately obvious);
- increase or to create short-term costs or negative consequences associated with the behaviour or action (that is, create disincentives to discourage the behaviour);
- reward or otherwise reinforce alternative behaviours or actions which do not have the long-term negative consequence (that is, offer incentives for change);

- regulate an activity, usually via a 'superordinate' authority (that is, put laws or some other mechanism in place to control the behaviour); and
- alter or avoid or reduce the long-term negative consequence (usually through technological means).

These kinds of solutions represent the broad range of possibilities open to the world for meeting the challenge of the issue of energy. After all, we are not addicted to energy as such, but certainly we are addicted to some of those things which abundant use of energy make possible. The aim of this book is to explore and further define some possible solutions: technological, political, behavioural and economic. The next chapter presents a summary history of energy supply and use, and a brief sketch of current patterns both globally and in Australia. Parts Two and Three discuss the technologies available in Australia and other countries that would allow us to achieve a sustainable energy system. In keeping with the recognition that energy is not simply a technological problem, these chapters also consider non-technical barriers; social, economic and political. The final two chapters in Part Four focus on policy measures and the means for reform.

References

Boyden, S. 1987. *Western civilization in biological perspective: patterns in biohistory*. Oxford: Clarendon Press.

Brechner, K.C. and Linder, D.E. 1981. A social trap analysis of energy distribution systems. In: Baum, A. and Singer, J. (eds), *Energy: psychological perspectives. Advances in Environmental Psychology* No. 3, pp. 27–51. Hillsdale, NJ: Lawrence Erlbaum Associates.

Goldemberg, J., Johansson, T., Reddy, A. and Williams, R. 1988. *Energy for a sustainable world*. New Delhi: Wiley Eastern.

Kapstein, E. 1990. *The insecure alliance: energy crises and western politics since 1944*. New York: Oxford University Press.

Kormondy, E. 1976. *Concepts of ecology*. 2nd ed. Englewood Cliffs, NJ: Prentice-Hall.

Odum, H.T. and Odum, E.C. 1976. *Energy basis for man and nature*. New York: McGraw-Hill.

Owens, S. 1986. *Energy, planning and urban form*. London: Prion.

Patterson, W.C. 1991. *The energy alternative: changing the way the world works*. London: Macdonald Optima.

Peet, J. 1992. *Energy and the ecological economics of sustainability*. Washington DC: Island Press.

Platt, J. 1973. Social traps. *American Psychologist*. 28: 641–651.

Ponting, C. 1992. *A green history of the world*. Harmondsworth, Middlesex: Penguin.

Slesser, M. (ed.). 1982. *Macmillan dictionary of energy*. London: Macmillan.

Tsai, H. 1989. *The energy illusion and economic stability: quantum causality*. New York: Praeger.

Wasi, P. 1991. An ecologically sustainable society: Buddhist perspectives. Paper presented to the Symposium, *Towards an Ecologically Sustainable Society: A Challenge for Community Education*, Centre for Resource and Environmental Studies, Australian National University, Canberra, 6–8 November 1991.

World Commission on Environment and Development (WCED). 1987. *Our common future*. Oxford: Oxford University Press.

CHAPTER 2

Historical and current patterns of energy use

STEPHEN DOVERS

This chapter is descriptive in nature, providing a backdrop for the succeeding chapters. It summarises the historical and current use of energy by the human population both globally and in Australia. The major problems arising from current energy systems will also be discussed. Finally, reference will be made to some predictions and extrapolations of future energy use, and to a selection of targets that have been presented as goals for achieving sustainable energy systems. The data here are generally presented in terms of the *energy* content of fuels (in giga- or peta-joules), rather than physical units such as litres or tonnes, thus facilitating comparison of different fuels.

Some qualifications apply to the data presented in this chapter and need to be kept in mind. The measurement and accounting of energy systems is a complex matter, and in many cases different countries, institutions and researchers use different conventions, terminologies and methods which can result in final numbers with significant variation. For example, some sources include fuelwood use, while others do not. Different sources may use different coefficients or factors to calculate the energy content of fuels, which often do in fact vary for fuels exploited in different geographic areas. Finally, there are variations in energy accounting conventions, such as in the case of hydro-electricity, with the energy value of this source being either calculated according to the actual energy output, or to the equivalent quantity of fossil fuels that would need to be used to produce that amount of energy (the latter giving a far higher number).

There is not space available here to comment further on the sources of error. The aim has been to keep the historical and current descriptions as simple and brief as possible. The data used necessarily contain

13

estimates and combinations of data from different sources. Thus some of the data given may vary according to source. For example, Australian national data will often be different from that taken from the standardised international data sets used for comparison across countries. None the less, the data can be regarded as representing the broad patterns and relative situations. The sources noted can be consulted by those interested in further detail.

Global overview

Historical perspective

As was stressed in the previous chapter, the development of modern societies has been underpinned, and indeed enabled, by the use of new forms and larger amounts of energy. Before considering the current global situation, it is useful to consider the longer historical perspective.

Table 2.1 provides estimates of human population and energy use from the time of the advent of farming some twelve thousand years ago up to the present. Obviously, the increase in total human energy use has been exponential and is still expanding. For most of human existence, the increase in energy use has been closely tied to population growth, but in recent centuries this nexus has been broken by huge increases in the exploitation of extrasomatic sources of energy. The most notable of these are coal, oil and gas, and their use is strongly

Table 2.1 *Estimated human energy use, 10,000 BC to 1990 AD*

Year	Population (billion)	Total energy use Somatic (PJ)	Extrasomatic (PJ)	Per capita extrasomatic (GJ/year)
(BC) 10,000	0.005	18	18	3.5
(AD) 0	0.25	900	1000	4
1650	0.53	1900	4000	7.5
1850	1.17	4250	11,500	10
1900	1.62	6000	29,000	18
1950	2.51	9250	72,000	30
1960	2.75	10,000	122,000	40
1970	3.70	13,500	200,000	55
1980	4.45	16,250	260,000	58
1990	5.30	19,400	360,000	67

Sources: Carr-Saunders (1936); United Nations (1976); Clark (1977); United Nations (1989); Boyden *et al.* (1990); British Petroleum (1991); Mannion (1991); author's estimates.
Note: The values in this table are approximations only, but serve to illustrate the pattern of change.

concentrated in the industrialised countries of the Northern Hemisphere. The Industrial Revolution in Europe and North America was the critical phase, a set of changes that have some similarity with the industrialisation processes currently in train and actively advanced in the developing world today. Spread around the current average per capita energy consumption rate of about 67 GJ/year, there is a range in different countries from less than 10 GJ to over 400 GJ.

In recent decades, there have been a number of important developments and events which have had a strong influence on patterns of energy consumption. The expansion of the world economy in the post-WWII period drove demand for energy ever upward. This prompted greater efforts in the proving and exploitation of energy resources, and this effort was increasingly expended in places other than the major energy-consuming industrial countries. Although present previous to this, the international trade in energy was greatly boosted and industrial societies in the Northern Hemisphere became more closely dependent upon imported energy. The advent of new fuels was also important. Oil was used in more applications and natural gas was exploited in larger quantities. Nuclear energy emerged from the wartime atomic weapons programs, and there was the prediction of the 'peaceful atom' producing electricity too cheap to meter.

This expanding picture has been disturbed by three factors. The first was the prospect of diminishing reserves and eventual energy scarcity. For some time, the standard response to this prospect was to further energy exploration activities. The second factor was somewhat more immediately startling, although not unrelated to the first. This was the 'oil crisis' of 1973, when the Organisation of Petroleum Exporting Countries (OPEC) quadrupled the price of crude oil on global markets, and sent much of the world into panic. The second oil shock of 1979 confirmed the lesson that energy supplies could not always be taken for granted, and that something should be done about this.

An unfortunate result of the timing of the oil shocks was that they came so soon after the idea of 'limits to growth' (Meadows *et al.* 1972) had been widely promoted. Whatever the faults in detail of the limits to growth arguments, at least the issues of finite resources and assimilative capacity of the environment were on the agenda. The oil price convulsions eventually stabilised and prices are once again low. To many commentators, this is the proof they desired that markets and the 'laws' of supply and demand would overcome any such resource and environmental problems. This interpretation misses two points. First, the oil shocks were not a resource scarcity issue, but rather a politically and economically imposed supply problem. Second, that running out of fossil fuels is one unwanted result, the other being the environmental consequences of continuing to burn them in vast quantities.

The net result of the oil shocks was the initiation of a good deal of discussion and analysis, and many initiatives in domestic energy production and in energy conservation. The growth in energy use slowed around the world, and actually decreased very briefly in some places. There was an increase in the effectiveness of energy use – that is, more goods and services were produced with the same amount of energy via efficiency gains (discussed later). The rate of increase in energy use has lifted again since the mid-1980s. The total consumption of commercial energy in the industrialised OECD countries over the last two decades is shown below (British Petroleum 1988; 1991), and a comparison of the 1980 and 1985 figures illustrates the impact of the oil shocks in the western world:

1970	135,000 PJ
1975	144,000
1980	157,000
1985	156,000
1990	174,000

Fuel shifts have occurred in recent years. Many countries have purposefully lessened their reliance on oil in response to the price rises of the 1970s. This has most often entailed an increased emphasis on natural gas or nuclear energy. Coal has continued to decline in relative importance, particularly in the industrialised world where its higher production of certain pollutants makes it unpopular.

The third disturbing factor relates to the environmental consequences of burning fossil fuels. For some time it has been apparent that the use of these fuels can cause local pollution, and some progress had been made in the control of these pollutants. It is only in the last few years, however, that the spectre of far greater possible consequences has arisen in the shape of climate change due to the enhanced greenhouse effect. Fossil fuel combustion is calculated to be the greatest contributor to this problem. This matter is discussed later in this chapter, but it is a fitting introduction to a consideration of current energy use that the world has so swiftly graduated from worrying about not having enough energy resources to worrying about what might happen if those available are used.

Current patterns

Broad current patterns of world primary energy consumption by major region and by fuel are summarised in Figure 2.1 and Table 2.2. Only a few major points require comment. There are large and obvious

INDUSTRIAL COUNTRIES

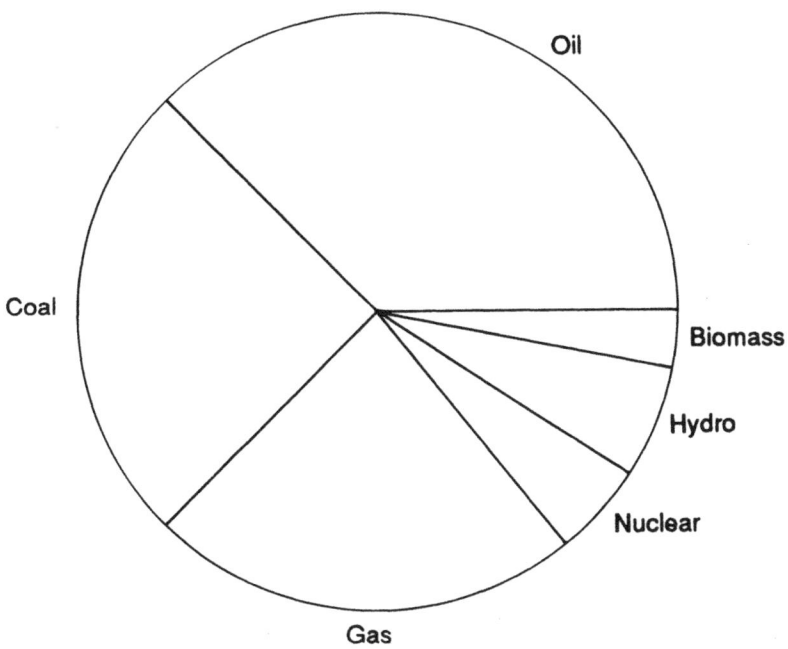

DEVELOPING COUNTRIES

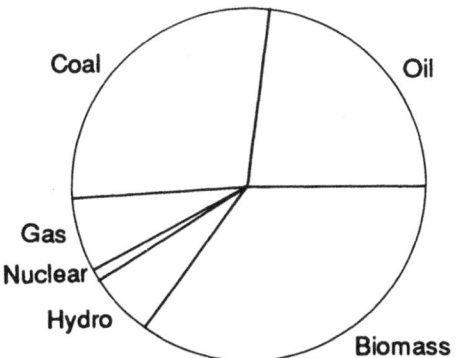

Figure 2.1 World primary energy use, 1990.

Table 2.2 *World primary energy use, 1990*

Region	Primary energy use Total (PJ)	Per capita (GJ)	Average annual growth (%) 1980–90	Oil	Natural Gas	Coal	Nuclear	Hydro	Traditional	Approximate somatic use (PJ)
Africa	13,700	21	4.2	29	10	23	1	6	31	2370
Asia	78,710	27	5.3	33	6	42	4	5	10	10,740
Middle East	10,250	78	7.5	61	37	1	—	1	0	480
USSR & Central Europe	71,590	186	1.2	28	37	26	4	4	1	1410
Western Europe	59,590	131	0.6	43	16	22	11	7	1	1670
North America	93,930	340	0.7	38	24	23	8	6	1	1010
Central and South America	23,320	52	3.4	46	14	4	0	20	16	1640
Australasia	4600	230	3.0	33	19	37	—	9	2	70
WORLD	355,690	67	1.9	37	21	26	5	6	5	19,400

Sources and Notes: Commercial fuels, and classification of regions, from British Petroleum (1991); traditional fuels (wood, etc.) from United Nations (1990) for 1988. Per capita and somatic energy based on United Nations (1989) medium variant population estimates for 1990. Australasia equals Australia and New Zealand; Western Europe includes eastern Germany and Turkey; USSR is pre-breakup; Japan is included in Asia; North America equals USA and Canada.

differences in the levels of energy consumption around the globe, and this North–South disparity is expanded upon in the next section. There are also large differences in the fuel mix used. For example, the proportion of regional totals contributed by wood and other traditional fuels ranges from near zero to almost a third, and for oil from less than 30% to over 60%. Within regional boundaries, there are of course significant country-to-country differences also.

The international trade in energy is highly significant, with numerous countries dependent on imports of energy to fuel economic activity, and others dependent upon the foreign currency earned through energy exports. In 1990, the oil trade ran at the level of about 24 million barrels per day (worth well over US$500 million/day at 1990 prices). The major oil importers are the USA, Western Europe and Japan (75% of total world imports). The major oil exporters are the Middle East (over 50% of total trade), along with Central and South America and northern and western Africa (British Petroleum 1991). Natural gas is an increasingly traded commodity, with pipelined exports totalling 233 billion cubic metres in 1990, and liquefied natural gas (LNG) 72 billion. The economic significance of energy trade to some countries is illustrated by the Australian case (with respect to oil, coal, LNG and uranium), which is discussed later. The countries of the world range from being awash with energy, to being energy-starved.

In terms of the availability of resources, the current situation with global energy reserves varies according to the fuel, and to how reserves are defined. *Proved* (or demonstrated economic) reserves are those which are known and could be extracted with existing technologies under prevailing economic conditions. *Sub-economic* reserves are known, but not economically viable to extract at present, and *inferred* reserves are those which can reasonably be expected to occur but have not been accurately demonstrated as yet. Obviously, resources tend only to be proved when a justification for this, usually economic, exists. *Proved* reserves of the major fossil fuels at the end of 1990 were (British Petroleum 1991):

- oil: 1009 billion barrels, being 43 years of production at current rates, with almost two-thirds located in the Middle East;
- natural gas: 119 trillion cubic metres, being 58 years of production at current rates, two-thirds located in the Middle East and the ex-USSR; and
- coal (all forms): 1079 billion tonnes, being 238 years of production at current rates, with 70% located in the USA, the ex-USSR, China and Australia.

Australia holds 0.2% of the world's proved reserves of oil, 0.4% of the proved reserves of natural gas, and 8.4% of coal. Apart from the 'absolute' constraint of the size of these non-renewable resource stocks, energy availability for any given society is also influenced by geographic location, economic factors (ability to buy or sell, or to exploit a resource), and political factors (of which the 1991 Gulf War provided a sharp illustration).

Distribution and disparities

Within the total global human energy budget there are enormous differences and disparities in the amounts of energy used, the sources utilised and the uses to which energy is put. A graphic way of illustrating these differences is to use the unit of a *human energy equivalent* (HEE), this being the amount of metabolic energy that flows through a human being, or about 3.6 GJ per year (Boyden *et al.* 1990). In the industrialised, high energy societies the per capita extrasomatic energy use typically ranges between 50 and 100 HEE – that is, as many as a hundred 'energy slaves' working for each woman, man and child in addition to their metabolic energy flow. At the other end of the scale some poor, developing countries have a per capita rate of only one or two HEE, roughly the same as that of a hunter–gatherer society.

The sources of this energy vary significantly, although the major fossil fuels of oil, coal and natural gas dominate overall. Wood remains a very significant fuel in the developing world, accounting for well over half of total energy supplies in some countries. But even though wood is theoretically a renewable energy source, there is a fuelwood crisis (Goldemberg *et al.* 1988). The sustainability of fuelwood supplies is a serious issue in many parts of the world, with demand and use clearly outstripping supply. This is not simply an energy issue, as removal of trees also relates to other issues such as soil erosion and nature conservation. Within the industrial world, some countries are less reliant on the major fossil fuels, either through utilisation of nuclear energy (e.g. France, western Germany, Sweden) or hydro (e.g. New Zealand, Norway, Canada) for a significant proportion of their electricity supply. Similarly, various countries rely to differing extents on particular fossil fuels, such as with a great reliance on either coal (e.g. China) or oil (e.g. much of the Middle East).

The implications of these differences are not restricted simply to energy supply and what this enables a person or a population to do, but also to the energy-related environmental impact of each person. For example, the global average per capita output of carbon dioxide, the major greenhouse gas, from industrial sources (mostly fossil fuel combustion) is about one tonne of carbon per year. The average for industrialised countries is over three tonnes, but for developing countries less than half a tonne (see Boyden and Dovers 1992: 64). When assessing these disparities, the differences between per capita and total somatic and extrasomatic energy use should be kept in mind. The following calculations for a few countries serve to illustrate this (energy data for 1988 from United Nations 1990; population data for 1990 from United Nations 1989; somatic energy assumed as 3.6 GJ/year):

United States of America:
Population	= 249 million
Extrasomatic energy use	= 325 GJ/capita/year
	= 80,925 PJ
Somatic energy use	= 896 PJ
Total human energy use	= 81,821 PJ

China:
Population	= 1135 million
Extrasomatic energy use	= 25 GJ/capita/year
	= 28,375 PJ
Somatic energy use	= 4086 PJ
Total human energy use	= 32,461 PJ

Australia:
Population	= 17 million
Extrasomatic energy use	= 216 GJ/capita/year
	= 3672 PJ
Somatic energy use	= 61 PJ
Total human energy use	= 3733 PJ

Japan:
Population	= 123 million
Extrasomatic energy use	= 134 GJ/capita/year
	= 16,482 PJ
Somatic energy use	= 443 PJ
Total human energy use	= 16,925 PJ

Papua New Guinea:
Population	= 4 million
Extrasomatic energy use	= 24 GJ/capita/year
	= 96 PJ
Somatic energy use	= 13 PJ
Total human energy use	= 109 PJ

The above calculations show that the impact of a small country (say, Australia) may be disproportionately large in relation to population due to a high per capita rate of resource consumption (for which energy use is a good proxy). Similarly, the impact of a large country with a low per capita consumption rate (say, China) is proportionately lower, although lifted by sheer weight of human numbers. When both the population and per capita consumption level are high, as in the case of the USA, a single country can become a very significant factor in the global equation.

A further variation in the pattern of world energy use relates to the effectiveness of the energy used in the economic system. This can be indicated by the *energy intensity* of an economy – that is, the quantity of energy (i.e. joules) used to produce a given economic output (i.e. Gross Domestic Product (GDP) in dollars). A feature of the period since the oil shocks of the 1970s has been a decrease, or at least a slowing in the growth of, the energy intensity of industrial economies. For the industrialised countries that make up the Organisation for Economic Cooperation and Development (OECD), energy intensity in 1973 was 22 GJ total primary energy supply per US$1000 GDP. In 1990, this had declined to 16 GJ/$1000, an average annual change of −1.7% (International Energy Agency 1991). This change can be attributed to a combination of structural change within economies, greater energy efficiency and fuel-switching. Between 1973–90, greater than average declines were enjoyed in some countries, such as the USA (from 24 to 18 GJ/$1000, or −1.9% per year) and Japan (16 to 11, or −2.4%). Much slower gains were made in Australia (21 to 20, or −0.6%), and in New Zealand energy intensity actually continued to increase.

As has been noted, energy is only a means to various ends. Energy intensity is a measure of the effectiveness of energy as a means to economic ends as measured by GDP. Figure 2.2 provides another, broader assessment. This graph plots per capita extrasomatic energy use for 126 countries against the United Nations Development Programme's (1990) *Human Development Index*. The Index combines measurements of life expectancy, adult literacy and adjusted GDP per capita. Thus, it is a useful broad measure of the achievement of human development or well-being. Compared as it is in this figure with per capita energy use, it gives an indication of the achievement of human well-being at varying levels of energy use (remembering the arguments put earlier that energy use can be a useful proxy for overall environment load). There are a number of qualifications that apply here. These include: the generalised nature of both data sets used; the impact of geographical and climatic factors on energy use; the impact of energy prices (e.g. in oil-producing states); and the impact of social, economic, health and education policies on human development. Despite these qualifications, the data do give a good broad picture of the relative situations.

At one level, the data presented in Figure 2.2 would confirm the belief that high-energy, industrialised societies are more able to achieve

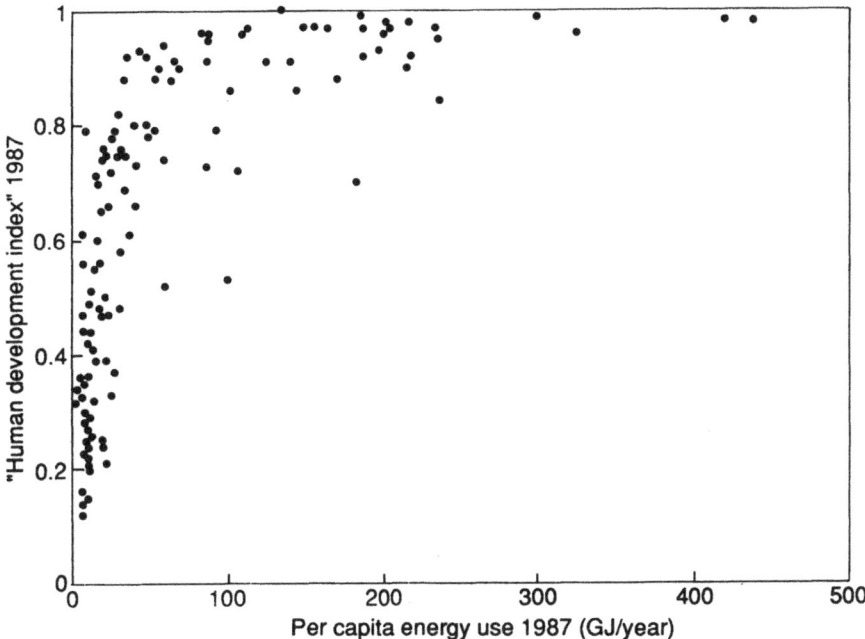

Figure 2.2 Energy use and human well-being.

higher levels of human well-being (assumedly their major end goal). However, some important and encouraging caveats can be applied to this interpretation with a closer examination of the data. These are:

- there appears to be a diminishing return operating, with the gains in terms of human well-being levelling off – that is, further additions in energy use (broadly reflecting resource consumption and industrial-isation) do not necessarily offer the same order of human develop-ment gains;
- some countries (those at the 'bend' of the curve) appear capable of delivering high levels of human well-being with comparatively low levels of energy/resource consumption;
- these countries are sufficiently diverse to apparently offer a range of social, political and economic settings under which such an achieve-ment might be made (although, from a climatic point of view, high latitude countries do not feature in this group).

This last point is emphasised by a sample of the countries that display a high Human Development Index (HDI) simultaneously with low GJ per capita energy use:

	GJ per capita	HDI
Cuba	64	0.877
Portugal	56	0.899
South Korea	69	0.903
Costa Rica	35	0.916
Hong Kong	59	0.936
Greece	87	0.949
Israel	83	0.957
Japan	134	0.996

These countries are comparable in their achievement of human development with ones who use from 150 to over 400 GJ of energy per capita. Among these countries are the USA (325 GJ/0.961 HDI) and Australia (216/0.978). It would appear that a high-energy, resource-intensive society is not necessarily a prerequisite for high levels of human well-being.

Australian overview

Australia is firmly placed within the high-energy group of countries. In fact it is near the top of this group, with an average per capita use of extrasomatic energy of over two hundred GJ per year. The transition to a high-energy society has been swift, with only two centuries having passed since European occupation of the country. Prior to that, the per capita energy use in Aboriginal Australia would have been 7–8 GJ per year, and the total energy flowing through the human system, both somatic and extrasomatic, somewhere between 3 and 7 PJ (depending on the actual population size, estimates of which vary). This energy comprised human metabolic energy, and fire used for cooking, warmth and as a land management tool.

Historical perspective

The European occupation of Australia in 1788 began a rapid transition from a low-energy hunter–gatherer culture to a high-energy industrialised one. This change can be tracked through a brief historical sketch of the human use of energy (for further information, see Corbett 1976; Saddler 1981).

European settlement in Australia began as the Industrial Revolution was accelerating in Britain, with coal assuming its critical dominance. However, somatic energy and fire continued as the main sources of energy for a few decades in the nascent colony, with forced convict labour particulary important. It is likely that the total human energy budget in Australia did not change markedly for a few decades. The swift reduction in the Aboriginal population caused by the European occupation by aggression, disease and displacement would have in part offset the initially slow increase in the white population. Soon draught animals became more and more important, especially in agricultural development, for transporting humans, for moving goods and materials, and for cultivation. This reliance on horses and bullocks necessitated the growing of large areas of fodder for the draught animals – a capturing of solar energy through crops and then through animals.

Although black coal was mined from the first few years of the 1800s, wood, human energy and draught animals remained the dominant forms of extrasomatic energy until later that century. Coal mining grew more quickly with the impetus of the gold rushes beginning in the 1850s. By this time steam-powered shipping had been introduced, and such industries as ironworks and the manufacture of town gas followed shortly. In the late 1800s, exports of coal began in earnest.

Australia was entering the industrial age, but draught animals and wood were none the less still the major sources of extrasomatic energy in rural areas and would remain so for decades to come. This was of some significance as this energy (especially horses) was enabling the agricultural development which produced the exports (especially wheat and wool) upon which the emerging national economy was to become reliant. Some oil products were available, most notably kerosene for lighting.

Railways, based on coal and steam, spread rapidly in the late nineteenth century, linking the major cities and ports with the primary production of their hinterlands, and also allowing the greater movement of people within those cities (which, for this among other reasons, spread and became of a lower density). Not insignificant also at around the same time was the appearance of the bicycle, which was a very good means of enhancing and making more effective human somatic energy. In the period of the 1880–90s, electricity generation began, usually based on a town generating plant.

The heyday of black coal, coming mostly from New South Wales, was the first two decades of this century. In the 1920s Victoria's brown coal was beginning to be exploited and oil, mostly imported, was being adopted as a fuel. Oil-powered motor transport was adopted rapidly, at first in the cities and later in rural areas. Electric tramways and electric power in factory motors also appeared around this time. Electricity supply was becoming more centralised, with the appearance of state electricity authorities and grid systems. Hydro-electricity was of considerable importance, most notably in Tasmania.

The production of black coal in this country topped 1000 PJ in 1875, rising above 10,000 PJ by 1915. By the Second World War, brown coal was approaching 5000 PJ, as was the amount of imported oil. Electricity generation, amounting to only a few hundred petajoules at the beginning of World War I, rose to over 3000 PJ by 1935 and more than double this by World War II. Following World War II, centralised electricity systems evolved rapidly and this was seen as a matter of national purpose. Saddler (1981) described this as the 'ideology of electrification', where electricity was seen as a symbol of progress and modernity. The Snowy Mountains Scheme, with its combination of hydro power and irrigation water, became an icon of national development. Oil gained in dominance at the expense of coal in many industries. In rural areas, the swift adoption of oil-powered transport and machines for cultivation and land clearance transformed both agriculture and the Australian landscape. From the 1950s, oil products such as fuel and diesel oil became cheaper and Australia's refinery capacity increased to virtual full capability. Coal's mainstay uses were to become that of feedstock for electricity generation and a few major industries (e.g. to make coke in iron and steel works), and as an export commodity.

Later, large oil reserves were discovered and domestic crude oil production began in the early 1960s. Natural gas was discovered in good quantities and exploited, becoming significant in the 1970s. Energy exports have become highly significant only in the last two decades, based on black coal and uranium. Energy use increased very quickly, driven by a combination of newly available fuels and technologies, economic expansion and population growth. Energy production and consumption and economic activity (as measured by GNP) have in recent decades displayed a close correlation, illustrating the vital nature of energy as an input throughout the economy. This energy growth has been a relatively recent phenomenon, but since the mid-1960s Australia has evolved the typical energy system of an industrialised society.

Current patterns

Energy use in Australia since World War II is summarised in Table 2.3, which shows clearly both the growth of energy use and the changes in the sources of energy. The current picture is shown in more detail in Table 2.4 (see later), which shows an energy balance for 1990–91.

Some terms require explanation at this point, which are basic to the data presented here and to energy budgets. *Production* of energy (or indigenous supply) includes all fuels produced in a country. With *exports* subtracted and *imports* added, this leaves the *total domestically available*

Table 2.3 *Summary of energy in Australia, 1950–90*

Year	Production, primary energy[A] (PJ)	Imports (PJ)	Exports (PJ)	FEC[B] (PJ)	Energy sources (%)[C]							Per capita (GJ/year)	
					1	2	3	4	5	6	7	FEC[D]	Domestic supply[D]
1950	890	280	3	na	21	3	18	48	2	8	—	c. 85	c. 100
1960	1270	630	120	940	10	2	10	63	4	11	—	90	
1970	2250	1050	610	1570	8	1	7	56	14	14	0	120	
1980–81	5260	620	2005	2150	7	1	7	52	17	16	0	145	
1984–85	7610	390	4390	2267								160	
1987–88	8100	500	5480	2480	7	1	7	50	18	17	0	150	
1990–91	10,300[E]	640	6840[E]	2780	7	1	6	49	19	18	0	165	235

Sources: Various, including United Nations (1976); Australia. Bureau of Resource Economics (1987); Australian Bureau of Agricultural and Resource Economics (1991).

Notes:
[A] Includes production for export.
[B] Final Energy Consumption; excludes exports and conversion losses.
[C] *1* = black coal and by-products; *2* = brown coal and briquettes; *3* = wood and bagasse; *4* = oil; *5* = natural and town gas; *6* = electricity; *7* = solar.
[D] Per capita FEC excludes exports and conversion losses; per capita domestic supply includes conversion losses.
[E] Inflated by anomalously high figure for uranium.

energy, or domestic supply, which is the total amount of *primary fuels* available for consumption within that country. Some of this is used directly (e.g. wood for heating, or hydro-electricity). However, much of it goes into various processes to produce *derived fuels*. Examples of this are coal, oil and gas being used to generate electricity, or crude oil being refined into automotive gasoline. This is called the *conversion sector*, and, in accord with the laws of energy, less energy emerges as derived fuels than goes in as primary fuels. The total primary and derived fuels available after the conversion processes is equal to the total *final energy consumption* of the various *end uses* of energy in the society.

Obviously, final energy consumption is significantly less than the sum of domestically available energy, the greatest reason being the loss of energy that occurs in generating electricity. Per capita *final* energy use in Australia is currently about 165 GJ per year, but per capita *total* energy use is about 235 GJ per year. (The latter figure is higher than that given earlier in the international comparisons, being calculated from more detailed, domestic sources.)

The last four decades have been characterised by strong and continued growth in the production and consumption of energy in Australia. Between 1973 and 1988, total demand for energy in Australia grew at the rate of 2.2% per year on average, higher than any other OECD country. This growth has been in both total and per capita figures, and in domestic use and total production including exports. The major factors in this growth have been as follows:

1.   Rates of population growth and economic growth in Australia have been consistently higher in recent years than in other industrialised countries. Energy demand is closely correlated to the size of the economy and population.
2.   Energy exports have risen dramatically in recent years. In the 1960s, a small amount of coal was exported. The energy content of these exports have increased more than six-fold in twenty years. The main exports are, in decreasing order of energy content, coal, uranium, oil products, and (most recently) natural gas.
3.   Over the same period, total imports have not changed significantly, and still comprise solely oil.
4.   New sources of energy have been exploited in the last few decades, these being:
     – indigenous production of natural gas in 1965 was negligible, but has risen to over 800 PJ in 1990, and now comprises almost 20% of final consumption;
     – centralised generation of electricity from fossil fuels has increased enormously. The final consumption of 72 PJ of electricity in

1960–61 has risen to almost 500 PJ in the early 1990s, with the relative contribution of hydro in this total having decreased a great deal; and

– uranium, although all exported and thus not used to produce energy within Australia, has, since the mid-1970s, become the second largest item in the Australian energy budget in energy terms.

An essential point to remember from this history is that both the rate of growth in energy use and the changes in the fuels used have occurred in the relatively recent past. This applies worldwide, but the full exploitation of particular fuels began somewhat later in Australia than in other industrialised countries. It would, therefore, be a serious mistake to assume that the current structure of the energy system is something that is given and will only change slowly. If history can be any guide here, then the lesson is that a nation's energy system can evolve and change very rapidly indeed. This is a possibility that has important implications for thinking about future energy systems.

The estimated 1990–91 energy budget in Table 2.4, along with the explanatory notes, provides a more detailed description of the current energy system. Some features, such as exports, have already been touched upon. It should be noted that the quantity of energy exported exceeds that used domestically. Part 2, conversion, illustrates the laws of energy well, with about three units of primary energy being used to produce one unit of derived fuel energy.

Of particular interest is part 4, describing which sectors use how much and what kinds of energy. End use is dominated by three fuels: oil, natural gas and electricity. Most oil, as automotive diesel oil and automotive gasoline (petrol), is used in transport. Most natural gas is used in industry, whereas electricity use is spread more widely amongst the various sectors. The industry and transport sectors dominate final energy use, accounting for over 70% of the total. In the transport sector, the proportional use of energy by different modes of transport is as follows (ABARE 1991):

Road   79%
Rail   3%
Air   12%
Water   6%

In the industrial sector, the two single largest users of energy are chemical (17%) and iron and steel industries (11%). In the residential sector, the bulk is used in space heating and water heating, which account for over two-thirds of energy used. Further detail of energy use in particular sectors is given in the Part Two of this book.

Table 2.4 (*Projected*) *Australian energy balance 1990–91 (petajoules)*

	Black coal	Coke	Coal by-products[D]	Brown coal[E]	Wood	Bagasse	Oil[F]	Gas[G]	Electricity	Solar	Uranium	Total
1. Energy supplies[A]												
production	4267.6			490.1	99.8	81.1	1285.7	874.2	57.7[H]	3.2	3142.0[I]	10,301.4
imports							638.4					638.4
exports	3063.2	29.6		0.9			430.2	178.4			3142.0[I]	6844.3
Total domestically available energy	1204.4	−29.6		489.2	99.8	81.1	1493.9	695.9	57.7	3.2		4095.8
2. Conversion sector[B]												
coke ovens	193.5	−119.3	−23.7				0.7		0.1			51.3
briquetting				0.1[J]					0.2			.0.3
petroleum refining								10.1	3.6			13.7
gas manufacture							1.6	−1.4				0.2
electricity generation	886.2		0.5	475.4			34.5	130.1	−494.1			1032.6
other conversion[G]		80.0[K]	−29.7				0.0		−20.7			29.7
fuel used in conversions							87.8	18.5	82.8			189.1
3. Energy available for domestic end use	124.7	9.7	52.9	13.8	99.8	81.1	1369.3	538.6	485.8	3.2		2778.9

Table 2.4 *continued*

	Black coal	Coke	Coal by-products[D]	Brown coal[E]	Wood	Bagasse	Oil[F]	Gas[G]	Electricity	Solar	Uranium	Total
4. End use												
agriculture							54.1		8.0			62.1
mining	6.1	0.3	1.2				37.8	92.2	36.3			173.9
industry, manufacturing, construction	110.3	9.4	51.7	11.5	23.0	81.1	159.4	321.1	198.2			965.7
transport	4.3						1037.1	1.3	6.8			1049.5
commercial	3.8			1.9	0.6		11.3	34.9	95.9			148.4
residential	0.3			0.3	76.2		16.2	89.1	140.5	3.2		325.8
lubricants, bitumen, etc.							53.4					53.4
Total final energy consumption	124.7	9.7	52.9	13.8	99.8	81.1	1369.3	538.6	485.8	3.2		2778.9

Source: Adapted from ABARE (1991).

Notes: (Minor discrepancies in totals are due to rounding.)

A Trade inclusive of stock changes and discrepancies.

B Negative figures in this section indicate amount of derived fuels produced via conversion processes.

C Includes blast furnace gas manufacture, cogeneration, etc.

D Coal tar, blast furnace gas, etc.

E Includes briquettes.

F Includes naturally occurring Liquified Petroleum Gas.

G Mostly natural gas; 0.5% town gas.

H Hydro-electricity.

I Unusually high figure due to stock changes and discrepancies; mean figure previous years *c.* 2000 PJ.

J 17 PJ brown coal used in briquetting to produce 16.9 PJ briquettes; 13.1 PJ briquettes and 0.7 PJ brown coal available for end use.

K Mostly coke used in blast furnaces.

This accounting is informative, except with respect to one important matter – what we actually use energy for. To know that a certain amount of energy is used in road transport is only part of the story, as it is just as interesting to know *what kinds of things are being transported*. Similarly, with industry it would be valuable to know which of the products we consume are the ones that require more energy to be manufactured. Industry does not use energy simply for the sake of it, but to produce some good or service which will be purchased and consumed by some individual, private firm or government body. What is required is another, supplementary form of energy accounting that shows energy use in terms of actual final consumption of goods and services.

A technique for doing this uses national *input–output tables*, which trace the flow of commodities, labour and capital between industries and thence to consumers. Fuels are one such commodity, and so energy can thus be allocated to the actual final use, known as *deliveries to final demand*. Final demand is the end point of consumption of a good or service. Such an exercise has been undertaken for Australia by Common and Salma (1992a, 1992b) who provide more detail and explain the methodology. Figure 2.3 shows energy intensity and energy use for deliveries to final demand for 1986–87 (data provided by Common and Salma, Centre for Resource and Environmental Studies, Canberra).

The twenty-seven sectors used are the most detailed breakup possible for these purposes, and represent categories of goods and services that are purchased by individuals or firms. The difference between this accounting and the energy balance used earlier can best be shown by examples. The energy used to make bricks and other building materials in this accounting is not allocated to the general 'industrial' sector, but rather to the category of *construction*, which better reflects the use we put those materials to, and the energy used to make them. A second example is in agriculture, where the significant energy input into fertilisers is allocated to the 'industry' sector in a standard energy balance. In Figure 2.3, the energy content of this and other inputs tc production are attributed to agriculture. A final example is the case ot transport, which is a single, not well-differentiated and very large category of end use of energy in the balance in Table 2.4. In final demand analysis, most of the energy used in transport is attributed to those areas of final consumption which use the transport to produce their goods and services for sale.

Some interesting perspectives emerge from a final demand analysis of energy. Energy intensity reflects the proportion that energy comprises of the total inputs of materials, energy, labour and capital required to produce goods or services in a sector. The three energy sectors shown (Petroleum and Coal Products, Electricity and Gas)

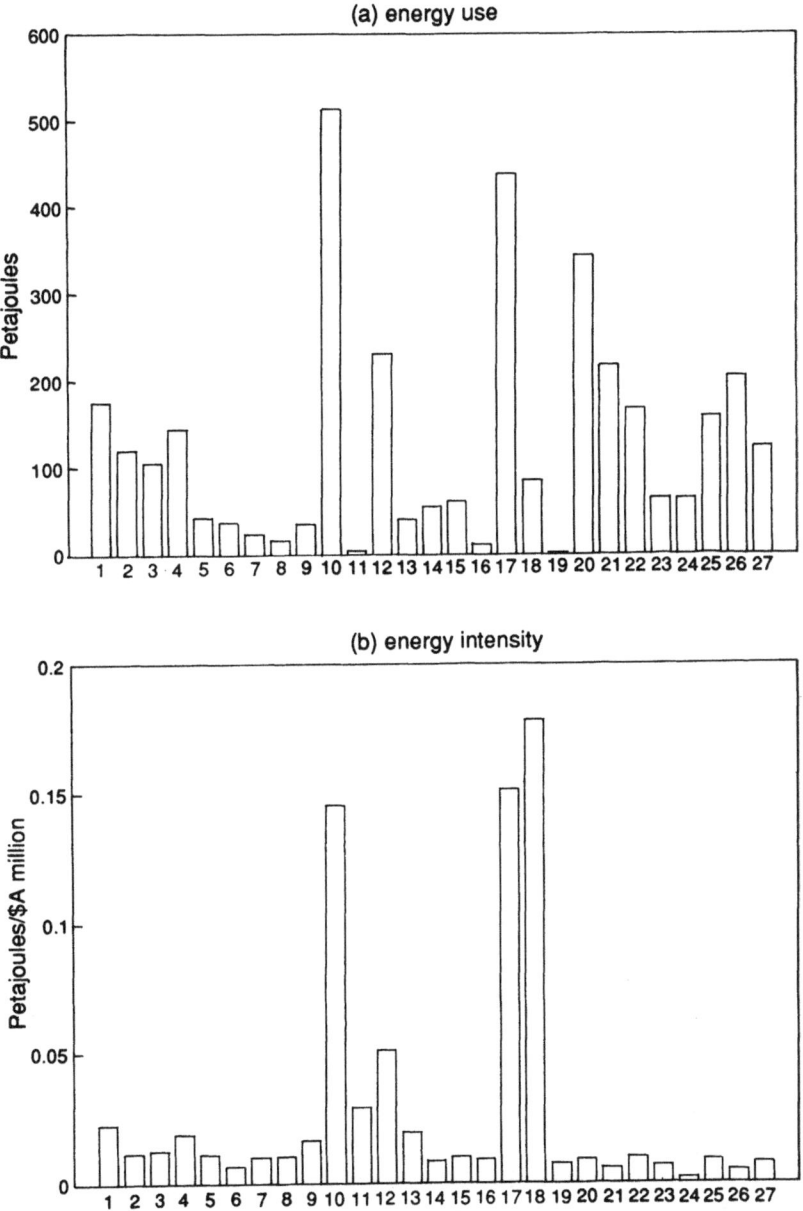

Figure 2.3 Energy requirements for deliveries to final demand, Australia, 1986–87.

have a very high energy intensity, which is to be expected as they produce energy-rich fuels as commodities for final consumption (e.g. petrol bought for use in cars, gas bought for heating). Other sectors which have a higher intensity include the two metals and minerals sectors of Non-metallic Mineral Products, and Basic Metals and Metal Products.

Total energy use is a function of energy intensity and the economic size of the sector. Petroleum and Coal Products (513 PJ) and Electricity (436 PJ) rank high, due to their very high energy intensity and their large size as economic sectors. Gas does not, reflecting the sector's smaller size. Basic Metals, a sizable industry and energy intensive through the requirement for heat, used 230 PJ. Some perhaps unexpected sectors also rank highly. Notable in this regard is Construction (343 PJ), with a low energy intensity but being a large sector. Also, the service and public sectors emerge as significant users of energy despite their low energy intensity, due again to their large size. Wholesale and Retail, Public Administration and Defence, Community Services, and Recreational and Personal Services together accounted for 704 PJ, or 20% of the total of 3476 PJ.

Herein lies the value of a final demand analysis as a supplement to traditional energy balance data. The use of energy is spread informatively across the many kinds of goods and services that a modern society such as Australia consumes. This belies any perception that energy is only used where there are engines running and furnaces burning. *All* sectors use energy through manufactured products, transport, heating, lighting, and so on.

Energy reserves

Energy reserves in Australia are, in total, large in relation to the size of the population. The Ecologically Sustainable Development (ESD) Working Groups (1991) provide the following life of demonstrated economic reserves at current production rates:

Black coal	330 years
Brown coal	880
Crude oil	13
Natural gas	45
Liquefied petroleum gas	30
Uranium	140

In addition there are quite large, at present sub-economic reserves of natural gas and shale oil, sizable inferred reserves of uranium, and vast inferred reserves of black coal. From the standpoint of supply of fossil fuels to the Australian economy, the more critical situation is that of oil.

In the absence of significant new oil finds, or diminished demand, the current high level of oil self-sufficiency will decline from the mid-1990s, necessitating a higher rate of imports. This will have significant economic and political implications. Regarding biomass energy as used in Australia, bagasse can be viewed as a fully renewable resource given the continuation of sugar cane growing, but one with little prospect of expansion. Fuelwood is theoretically renewable, but is at present a matter of concern in relation to impacts on native vegetation and wild-life habitat. Australia's forest resources are limited, so a conscious societal decision to expand the utilisation of wood as fuel significantly and sustainably would need to be based upon fuelwood or mixed-purpose plantations. Further prospects for hydro-electricity in Australia (i.e. the 'reserves' of suitable sites) are constrained by hydrological, geographical, economic and environmental factors (except in a few areas such as parts of Tasmania).

This consideration of energy production and consumption in modern Australia has so far omitted two significant sources of energy. (Also missing from the analysis is the contribution of draught animals, but this is negligible.) The first is human somatic energy. Although, at around 60–65 PJ, it is minor in terms of the total energy budget, it is of course absolutely indispensable, and important in the context of human health and well-being.

The second is solar energy. The energy balance used here (Table 2.4) shows solar energy accounting for a mere 0.1% of the total domestically available energy. However, this amount is that captured via roof-top collectors in the residential sector, and obviously we use solar energy in many other ways. But these are not energy sources that pass through markets or are measured in any other way, such as drying clothes, growing and drying crops, warming houses, and so on. Kaneff (1991) provides some estimates of such uses of solar energy, and claims that, if included in an energy budget, solar energy used in salt production would account for around 14% and food production a further 6.5% of the total primary energy supply. Such figures make solar energy somewhat more significant.

It is estimated that the solar energy falling on the Australian continent is the equivalent of over twelve thousand times the annual demand for the major fossil fuels in the country (ESDWG 1991). They also noted that effective solar collectors on every rooftop could provide more than the total electricity consumption of the country. Solar is thus an enormous potential resource. There is also potential for other forms of renewable energy at present untapped or only marginally utilised (wind, biomass, tidal, etc.), subject to technical, economic and locational constraints. These potentials are explored further in Part Three of this book.

Economic significance

There are a number of ways of assessing the economic significance of energy in Australia. From one viewpoint energy is absolutely essential, given that in its absence all production and consumption would cease. From another extreme, energy is simply one of many commodities that are bought and sold in an economic system and thus is simply a commodity that is subject to the same pressures of supply, demand and substitution as all others. A realistic view lies somewhere between.

Energy certainly is a special case as an integrative and indispensable input to production, as has been made clear earlier. Even disregarding this, the energy sector is economically very important in Australia. In a nation heavily reliant upon commodity export earnings, energy exports account for almost a fifth of the value of total commodity exports. The energy sector also accounts for approximately 5% of Gross Domestic Product and some 2% of total employment.

The relative energy requirements of different sectors of the economy vary, as the final demand analysis reported on above indicates. The ESD Working Groups (1991) give the proportion of total intermediate inputs attributable to expenditure on energy for a number of sectors as follows:

Farming	9%
Mining	11%
Manufacturing	12%
Transport	21%
Commercial	9%
Residential	3%

Energy is thus an input of some relative significance in many parts of the economy (noting that personal transport is not included in the residential sector here). The output of goods and services per unit of energy input has increased in recent years in Australia – that is, the energy intensity of the economy has decreased, as noted above. The decline in energy intensity in this country has been less than in other OECD countries.

Finally, the economic significance of energy may assume new proportions in the future. An increasing dependence upon oil imports would have repercussions for the national trade balance, as would any international moves either to or away from uranium or coal as fuels for environmental or other reasons.

Problems

Increasing energy use has had many beneficial outcomes – in essence it has allowed many people and societies to *do more*. It has, however, brought with it many serious problems. The more critical ones are noted in this section.

An important set of considerations to be kept in mind here are the differences in space and time scales of the various problems. Some may be at least potentially short-lived and localised in nature, such as some cases of pollution, urban congestion or environmental disturbance. At the other end of the scale, climate change is a long-term and completely global problem, although in effect the sum of many individual or national activities. In between there are a range of issues. It is the long-term and international issues which are more intractable in terms of reform, which are dominating the current global environmental agenda, and which have generally been recognised more recently. They are also ones which strongly implicate energy systems.

Air pollution

When matter is combusted, various solid and gaseous by-products are produced. Fuels are no exception, and the emissions from them are diverse. Air pollution problems from energy sources date back to the first use of fire in enclosed spaces millenia ago. (Such problems persist in fuelwood-dependent countries; the World Bank (1992) estimates between 300–700 million people, mostly women and children, being adversely affected by indoor smoke.) At a societal level, centuries of concern but inactivity in Britain preceded the dreadful smog episode of 1952 (four thousand people killed) that led the government to finally controlling the burning of coal in open fires. Around the world today air pollution problems are legion, and the major cities and industrial areas of Australia are no exception.

Today, the bulk of air pollution problems are rooted in the energy sector (Evans 1989; French 1990). Key parts of the energy production and consumption sector and some major pollutants produced in them are:

- motor vehicles: carbon monoxide, lead, nitrogen oxides, hydrocarbons;
- fossil fuel-fired electricity generation: sulfur dioxide, nitrogen oxides, particulates;
- fuel processing (e.g. refineries, coke ovens): carbon monoxide, sulfur dioxide, hydrocarbons.

In Australia, data presented in Farrington (1988) for the airsheds of major Australian cities indicates that fuel use in transport and fuel combustion elsewhere accounts for 95% of the carbon monoxide, 57% of the hydrocarbons, 91% of nitrogen oxides and 55% of sulfur dioxide emitted. This does not include other minor energy sources, such as utility motors, petroleum evaporation or gas leaks. By far the greatest single source of urban air pollution in developed countries like Australia is the motor vehicle.

In addition, secondary air pollutants are produced by the chemical reaction of different primary pollutants. Of particular importance from a health viewpoint are photochemical smog and ozone. Acid precipitation ('acid rain') follows atmospheric reactions and subsequent leaching involving oxides of sulfur and nitrogen. Although not judged yet to be a problem in Australia except in some specific locations, it is a matter of great concern in the Northern Hemisphere, for example in having caused damage to over a third of the forests of Europe (Australian Environment Council 1989).

The direct and indirect impacts of air pollutants are a worldwide concern, despite some degree of success in controlling emissions. Some major impacts are:

- direct human health impacts, usually being the result of breathing in pollutants (air quality standards are regularly exceeded in cities and industrial areas throughout all regions of the world; in some cities, breathing the air has been compared to smoking ten cigarettes a day);
- impacts on biophysical production, such as soils, wetlands and fisheries, forests, etc.;
- impacts on buildings, monuments and infrastructure, such as statues, historic facades, exposed metals, water pipes, etc.

The enhanced greenhouse effect

In addition to the various emissions summarised above, the combustion of fossil fuels has another and quite different impact on the atmosphere. The natural presence of a number of 'greenhouse gases' in the atmosphere maintains the temperature of the globe much higher than would otherwise be the case (and all the things that flow from that, including the whole climatic system). These gases include water vapour and carbon dioxide. This phenomenon has been well-established scientifically, and is known as the *greenhouse effect*.

Essentially since the Industrial Revolution, humans have increasingly caused the emission of various greenhouse gases, thus enhancing (increasing) the natural greenhouse effect. The causes and possible impacts of this are well-covered elsewhere (Pearman 1988; Henderson-Sellers and Blong 1989; Flavin 1990; Intergovernmental Panel on Climate Change 1990; Industry Commission 1991), and need be only briefly noted here.

Although the actual predictions of what will happen are subject to great uncertainty, there is a reasonable consensus among scientists of the effects at the global level. Mean global temperatures are expected to rise (2–5°C with a doubling of atmospheric carbon dioxide), rainfall patterns are expected to change, and sea levels are expected to rise. At regional or national scales the predictions are very uncertain, but the implications of the possible changes are considered by many to be very serious. These include inundation of human settlements, changed and often worsened occurrences of natural disasters, and severely dislocated patterns of primary production.

The main human-influenced greenhouse gases, their main sources and their relative contribution to global warming are:

- carbon dioxide: fuel combustion, biomass burning; responsible for over 50% of predicted global warming;
- methane: farming, biomass burning, fuels and industry; 10–20% of warming potential;
- nitrogen oxides: agriculture, fuels and industry; 5–10%; and
- chlorofluorocarbons: industry (synthetic molecules); 15–20%.

Before large-scale industrialisation, the concentration of carbon dioxide in the atmosphere was about 280 parts per million (ppm). In 1990 it was over 350 ppm and growing by about 1.5 ppm per year. Global carbon dioxide emissions at the start of the 1990s were over 6000 million tonnes carbon content. From WWII until the early 1970s, emissions grew at a rate of over 4% per year. Growth rates slowed between then and the mid-1980s, but have since moved upwards again.

The bulk of carbon dioxide emission comes from the combustion of fossil fuels. Energy sources also contribute to emissions of methane and oxides of nitrogen. Taken together, emissions from energy systems are the largest single contributor to the enhanced greenhouse effect. This fact makes energy systems an obvious target for addressing the threat of climate change. Indeed, reductions in energy use and the possibility of utilising non-carbon fuels has been the major focus of international and national discussion regarding climate change.

Australian carbon dioxide emissions are high on a per capita basis, as would be expected for an energy-intensive society, and are by far the main greenhouse gas source arising from this country. Estimates of emissions vary due to the existence of different factors used to extrapolate emissions from energy use (this lack of consensus is a matter of concern, as the variations resulting from different sets of factors approach 10%). Different fuels emit different amounts of carbon per unit of energy gained when burnt. Of the main fuels, brown coal emits the most carbon, then black coal, oil, wood and bagasse, and natural gas the least. On the basis of some factors, natural gas emits half the amount of carbon than does brown coal for the same energy yield. The Australian Institute of Petroleum (AIP 1989) gives the CO_2 intensity of major fuels as: black coal 104 kilotonnes CO_2/petajoule; brown coal 113; wood and bagasse 79; oil 73; and gas 55 kt/PJ. Solar and nuclear energy systems do not emit carbon during operation. Electricity's carbon intensity varies with the fuel (coal, gas, oil), and with the efficiency of the generation process. Its contribution can be included by calculating total emissions prior to energy conversion processes. As a rough guide, a PJ of electricity consumed in Australia will equal some 300 kt of CO_2.

Common and Salma (1992a), using the AIP factors, calculated 217 million tonnes of carbon dioxide from energy use in Australia for 1973–74, rising to 321 mt for 1988–89, a growth rate of 2.5% per annum. On this basis, for the total domestically available energy of 4096 PJ from Table 2.4, the projected CO_2 emission total for 1990–91 would be 342 mt. ABARE (1991) calculated 280 mt for 1989–90. CO_2 emissions follow energy use quite closely. On the basis of a standard energy balance, the transport and industry sectors account for the bulk of emissions. The final demand analysis of Common and Salma (1992a, 1992b) also calculates CO_2 for 1986–87 arising from energy use in the twenty-seven sectors. From their 1986–87 total CO_2 emission of 292 mt, the five major contributing sectors are:

- Electricity   15%
- Petroleum and Coal Products 13%
- Construction   10%
- Basic Metals   7%
- Wholesale and Retail   6%

Total Australian emissions will continue to rise in line with increases in fossil fuel use. If we assume a round figure of 300 million tonnes, then the per capita figure is 17 or 18 tonnes of CO_2 per year – about four times the world average. However, the total is only a little over 1% of the world total. These figures do not take into account the carbon content of Australian coal exports, which is only a little less than that of all fuels combusted domestically (uranium exports do not contain carbon).

Other problems

High or unsustainable levels of energy use have lead to numerous other problems, a few of which warrant mention here. Issues of resource supply, covered earlier, are obviously relevant. The energy systems in industrialised countries is such that an absolute reliance on motor vehicles has emerged in many societies. This has shaped the structure and function of our cities, with a close relationship developing between high energy use and low urban density (Newman and Kenworthy 1989; see also Chapter 4). Australian cities are among the most energy-wasteful and sprawling in the world. The problems of motor vehicle dependency include traffic congestion, road accidents and urban alienation and dislocation.

The extraction and processing of fuels involves problems other than the air pollutants discussed. The problem of nuclear wastes is a particularly problematic one. Also, much energy resource exploration and extraction can cause on-site disturbance – for instance the impacts of coal mines and fuel pipelines. Oil spills are a particularly notorious form of pollution.

Apart from a fuelwood shortage, the use of fuelwood in many parts of the world has led to land degradation (especially soil erosion), degraded water catchments, changed runoff patterns and the loss of nature conservation values. The use of dung and crop residues for fuel removes valuable organic matter and nutrients from the agricultural system, either impoverishing soils or necessitating the input of artificial fertilisers to maintain productivity.

Future prospects

To any observer of current debates around the world concerning environment and development, the fact that our current energy systems are under question is glaringly obvious. A fair consensus exists that the world cannot continue to use more energy, at least not those forms we use now. The energy sectors themselves, governments, industry and the general community – all are to some degree thinking about the future of energy systems. Some are starting to act. One detail that is missing is a clear idea of where we should be heading: how much less energy, and by when. There is in fact very little agreement as to what degree of change is required, partly because even the near future is clouded with uncertainty. This short section discusses some projections of energy use, and notes a small number of suggested targets for energy system reform.

Energy use is closely tied to population growth and economic growth, as well as to technology and resource availability. Population growth will continue globally for decades to come, in fact the 1990s will see a greater increment added to the human population than ever before. The world population is expected to stabilise sometime next century at a level anywhere between eight and fourteen billion. Most of this growth will happen in developing countries which have a low per capita rate of energy use, but where this rate is rising. The World Commission on Environment and Development (1987) noted that if the energy- and resource-intensity of the developing world was to become equal to that of the developed world, then the 'load' on the biosphere would increase five-fold. It also noted that the biosphere could not withstand such an increase. Factoring in population growth over time makes this scenario even more grim. Future population growth is less uncertain than future economic growth, upon which per capita energy use greatly depends. Projecting economic growth rates into the near term is a brave enough art, let alone over half a century or more. Similarly, technological development is unpredictable. Rapid development of technologies such as viable nuclear fusion or solar–hydrogen fuel systems would change the energy future considerably.

In spite of these difficulties there are those courageous enough to predict future patterns of world energy use. Goldemberg *et al.* (1988: 14) has summarised a number of these, and the range is startling. Some predictions include a degree of conscious change in response to climate change, and these project a lower or at least not too greatly increased level of energy use. Others assume a continuation of present population and economic trends, and range as high as a four-fold increase in global energy use by 2030. On the basis of the range of predictions available, the 1990 world energy use total of about 360,000 PJ might reasonably be expected to double in as little as the next quarter century in the absence of moves to stop this happening. However, it would appear likely that moves to limit the growth of energy use will be pursued to some, as yet unclear, extent. Rational projections beyond the next two or three decades do not exist for understandable reasons.

The Australian Bureau of Agricultural and Resource Economics (ABARE 1991) has projected Australian energy demand and supply forward to 2004–05. These projections are given below, with the corresponding 1989–90 figure of fifteen years previously given in parentheses for comparison:

- indigenous production: 13,961 PJ (8922)
- imports: 1087 PJ (625)
- exports: 9668 PJ (5250)
- total domestically available energy: 5381 PJ (3948)
- final energy consumption: 3669 PJ (2664)

There is obviously an important difference between what even adjusted extrapolations would suggest energy use will be, and what various people believe it should be. A number of groups have suggested targets for future energy use, being responses to environmental issues and in particular climate change. An international meeting in 1988 agreed to a target for the reduction of greenhouse gas emissions (Conference on the Changing Atmosphere 1988). This target has been accepted by some countries and rejected by others. The Australian Government has accepted this as the basis for its interim planning target – to reduce greenhouse gas emissions by 20% by the year 2005, from a base of 1988 levels (subject to caveats regarding international economic competitiveness). Should this target be met fully and only through the energy sector, the reductions in energy use would of course be more than 20%. A number of players in the debate (such as some western governments and many industry groups) state that this is an unreasonably severe target; others state that it is too low.

The Intergovernmental Panel on Climate Change (IPCC 1990) stated that, to stabilise atmospheric carbon dioxide concentrations at the present, already elevated level, reductions in the human-made emissions of this gas would need to be 60–80%. For such stabilisation of other greenhouse gases, the reductions required are calculated to be of a similar order as this, except for methane which is only 15–20%. As there are a number of sources and potential sinks of greenhouse gases, the range of possible reduction or stabilisation strategies is large. However, given that energy sources are the largest, the 60–80% reduction figure provides a good idea of the magnitude of the problem based on the best available knowledge.

Ignoring population growth, and assuming the goals of keeping gross world resource consumption at current levels and achieving equality in consumption, a global mean per capita energy consumption rate of around 70 GJ/year has been put forward by Boyden et al. (1990). This is about the same as the current world average. They further propose that renewable (especially solar) energy systems become increasingly important. Such a target would involve increases in many parts of the developing world and varying decreases throughout the industrialised world depending on the present level of energy use. For Australia, this would mean a reduction to about one-third of the current per capita level (to the same level as about four decades ago). Trainer (1991: 20) has suggested per capita energy use in the rich world one-tenth of that consumed presently as sustainable, noting that this would require 'quite radical change ... in lifestyles, the economy and in the pattern of settlement'.

Goldemberg et al. (1988) have suggested that the energy consumption of industrialised countries could be cut by half within the next three decades, and that, at the same time, increases in energy use in the developing world would result in a global energy use total only slightly higher than that at present. They further state that this could be achieved with continued economic growth and improved human well-being. In their words, 'the future of energy is much more a matter of choice than of prediction'.

Even on the basis of this sketchy survey, suggested goals for an energy future range from little change from the present situation to massive cuts. The targets are very approximate, and many obviously would see some of them as ridiculous. But two things emerge from considering such goals. The first is that consideration of what the goal should actually be must surely be an area needing urgent attention. The second is that, whatever the eventual consensus is, the change apparently required in energy systems will be profound enough to strongly justify purposeful action in the short term.

A note on nuclear energy

This book does not consider nuclear energy apart from its contribution to current energy use as described in this chapter, and minor comments elsewhere. This exclusion does not imply a belief that nuclear energy either will not or cannot play a role in the energy systems of the future – in all likelihood it will. The exclusion of nuclear energy is made on the basis of three considerations. The first is simply that, although Australia is a supplier of uranium to other countries, nuclear energy is not used in this country, and at present this is official government policy. It is also government policy at the time of writing that uranium production be limited to the output of three mines only.

The second consideration is that the focus of the book is on conservation of energy and renewable energy sources. Nuclear energy is not a renewable energy source, being based on an eventually exhaustible stock of minerals. In future, technological developments (fusion and breeder reactors) may alter this status, but these technologies are apparently some distance from being viable. With respect to energy conservation, reduced demand or increased efficiency at end use point will affect all existing energy sources, nuclear or otherwise, in the same manner.

The third and main consideration is the presence of a number of unsolved problems and questions which hang over this energy source. The World Commission on Environment and Development summarised the situation by stating that (WCED 1987: 189):

> The generation of nuclear power is only justifiable if there are solid solutions to the presently unsolved problems to which it gives rise.

These issues are discussed at length elsewhere, and can be merely noted briefly here (see Flavin 1987; WCED 1987: 181–189). To summarise, the unsolved problems can be grouped into three areas:

- health and environmental risks from radioactive material (including leakages in operation, the risk of major accidents, and the problem of disposal and storage of the very long-lived wastes);
- the problematic and controversial linkages to nuclear weapons proliferation; and
- problems of cost and scale: nuclear power plants are a large and costly technology, they are by definition a centralised energy source, and the costs associated with decommissioning plants at the end of their operative life are only just beginning to be assessed.

It should be noted that these problems persist in spite of massive research and development expenditure on nuclear energy; the World Bank (1992: 19) estimated that sixty per cent of public research funds for energy are allocated to nuclear, compared to six per cent for renewable energy research.

Conclusion

This chapter has provided a brief sketch of the importance of the issue of energy, the historical and current patterns of use, and the major problems associated with the present energy systems in place around the globe. While this has only been a sketch, it should have established the basic assumptions of this book. These are that:

- the issue of energy is fundamental to any discussion addressing the challenge of long-term ecological and economic sustainability;
- extant energy systems are on the whole not ecologically sustainable in the long term; and
- a major challenge is therefore to actively seek to achieve new energy systems that are far more ecologically sustainable, and which can offer the means of achieving and maintaining the reasonable health and well-being needs of all people.

The magnitude of this challenge should not be understated, but nor should the potential rewards. The means to address the challenge will involve technological, political, social and economic change. The remainder of this book explores these.

There are two facets to this challenge. The first can be loosely termed *energy conservation* – the reduction of the amount of energy required to produce required goods or services. In terms of where improvements in efficiency and reductions in use can be made, the current pattern of energy use would suggest buildings, industry and transport as priority areas. The next part of the book deals with what may be possible in these sectors. The second facet of the challenge is to exploit *new or alternative sources of energy* that avoid or minimise the environmental problems associated with the present fossil-fuel-based system. Ideally, these will be sources of renewable energy. Part Three of the book surveys the state of play in this area. Part Four concentrates on the policies that might be used to enable reform in the energy system.

References

Australia. Bureau of Resource Economics. 1987. *Energy demand and supply, Australia, 1960–61 to 1984–85.* Canberra: Australian Government Publishing Service.

Australian Bureau of Agricultural and Resource Economics (ABARE). 1991. *Projections of energy demand and supply, Australia, 1990–91 to 2004–05.* Canberra: Australian Government Publishing Service.

Australian Environment Council (AEC). 1989. *Acid rain in Australia: a national assessment.* AEC report 25. Canberra: Australian Government Publishing Service.

Australian Institute of Petroleum (AIP). 1989. *The greenhouse effect: a position paper.* Melbourne: AIP.

Boyden, S. and Dovers, S. 1992. Natural-resource consumption and its environmental impacts in the western world: impacts of increasing per capita consumption. *Ambio.* 21 (1): 63–69.

Boyden, S., Dovers, S. and Shirlow, M. 1990. *Our biosphere under threat: ecological realities and Australia's opportunities.* Melbourne: Oxford University Press.

British Petroleum (BP). 1988. *BP statistical review of world energy 1988.* London: BP.

British Petroleum (BP). 1991. *BP statistical review of world energy 1991.* London: BP.

Carr-Saunders, A.M. 1936. *World population: past growth and present trends.* Oxford: Clarendon Press.

Clark, C. 1977. *Population growth and land use.* 2nd ed. London: Macmillan.

Common, M. and Salma, U. 1992a. *An economic analysis of Australian carbon dioxide emissions and energy use.* Report to the Energy Research and Development Corporation. Canberra: Centre for Resource and Environmental Studies, Australian National University.

Common, M.S. and Salma, U. 1992b. Accounting for changes in Australian carbon dioxide emissions. *Energy Policy.* 20: 217–225.

Conference on the Changing Atmosphere: Implications for Global Security. 1988. Conference statement. Toronto, 27–30 June, 1988.

Corbett, A.H. 1976. *Energy for Australia: resources, technology and the environment.* Ringwood, Victoria: Penguin.

Ecologically Sustainable Development (ESD) Working Groups. 1991. *Final report – energy production,* and *Final report – energy use.* Canberra: Australian Government Publishing Service.

Evans, D. 1989. Emission and control of air pollution from energy industries. In: Jakeman, A. (ed.), *Air pollutants from energy industries: scientific basis of standards and research needs*, pp. 1–9. Canberra: Centre for Resource and Environmental Studies, Australian National University.

Farrington, V. 1988. *Air emission inventories (1985) for the Australian capital cities.* Australin Environment Council report 22. Canberra: Australian Government Publishing Service.

Flavin, C. 1987. *Reassessing nuclear power: the fallout from Chernobyl.* Worldwatch paper 75. Washington DC: Worldwatch Institute.

Flavin, C. 1990. Slowing global warming. In: Brown, L. (ed.), *State of the world 1990*, pp. 17–38. New York: W.W. Norton.

French, H. 1990. Clearing the air. In: Brown, L. (ed.), *State of the world 1990*, pp. 98–118. New York: W.W. Norton.

Goldemberg, J., Johansson, T., Reddy, A. and Williams, R. 1988. *Energy for a sustainable world.* New Delhi: Wiley Eastern.

Henderson-Sellers, A. and Blong, R. 1989. *The greenhouse effect: living in a warmer Australia.* Sydney: University of New South Wales Press.

Industry Commission. 1991. *Costs and benefits of reducing greenhouse gas emissions.* 2 volumes. Canberra: Australian Government Publishing Service.

Intergovernmental Panel on Climate Change. 1990. *Climate change: the IPCC scientific assessment.* Cambridge: Cambridge University Press.

International Energy Agency. 1991. *Energy policies of IEA countries: 1990 review.* Paris: Organisation for Economic Cooperation and Development.

Kaneff, S. 1991. Renewable energy. In: Dovers, S. (ed.), *Energy options for sustainability*, pp. 48–100. Canberra: Centre for Resource and Environmental Studies, Australian National University.

Mannion, A.M. 1991. *Global environmental change: a natural and cultural environmental history.* Harlow, Essex: Longman.

Meadows, D.H., Meadows, D.L., Randers, J. and Behrens, W. 1972. *The limits to growth.* New York: Earth Island.

Newman, P. and Kenworthy, J. 1989. *Cities and automobile dependence: an international sourcebook.* Aldershot: Gower.

Pearman, G. (ed.) 1988. *Greenhouse: planning for climate change.* Melbourne: CSIRO/Brill.

Saddler, H. 1981. *Energy in Australia: politics and economics.* North Sydney: George Allen and Unwin.

Trainer, T. 1991. Thinking about the nature of the required conserver society. In: Cock, P. (ed.), *Social structures for sustainability*, pp. 19–26. Canberra: Centre for Resource and Environmental Studies, Australian National University.

United Nations. 1976. *World energy supplies 1950–74.* New York: UN.

United Nations. 1989. *World population prospects 1988.* New York: UN.

United Nations. 1990. *Energy statistics yearbook 1988.* New York: UN.

United Nations Development Programme. 1990. *Human development report 1990.* New York and Oxford: Oxford University Press.

World Bank. 1992. *World development report 1992: development and the environment.* New York: Oxford University Press.

World Commission on Environment and Development (WCED). 1987. *Our common future.* Oxford: Oxford University Press.

PART TWO

Efficiency and conservation

CHAPTER 3

Using energy efficiently in buildings and industry

HUGH SADDLER

As previous chapters have demonstrated, concern about the sustainability of energy systems is not a recent phenomenon. The exhaustion, actual or potential, of fuel resources and the pollution caused by fuel combustion have affected some human societies for centuries.

It is, however, only in the past two decades that these twin concerns have assumed global significance and become major topics of national and international policy debate. The so-called first oil shock, when OPEC countries combined in 1973 to increase the price of oil fourfold, marks a useful starting point for the emergence of this level of concern. Since then, the rate of production of papers, books, reports, official statements and policies on energy technology, economics and policy has increased enormously. Most have sought to address issues related, in one way or another, to either pollution or resource depletion.

To simplify greatly, responses to these problems may take one of two approaches:

1. A supply-oriented approach seeks to solve problems of resource depletion by moving to new energy sources, e.g. from oil to nuclear fission or solar energy, and to solve problems of pollution by the use of technical measures to reduce polluting emissions.
2. A demand-oriented approach, commonly termed energy conservation, seeks to ameliorate both problems by reducing the total quantity of energy used by society.

Of course, the two approaches are not mutually exclusive, and most comprehensive energy policy documents produced during the last twenty years have claimed to embody an appropriate blend of both. However, to a dispassionate reader of these documents, the great majority of them are oriented predominantly towards one or the other approach.

Such a reader would also notice that over the past twenty years there has been a decisive shift in the consensus as to which approach should predominate. Twenty years ago pronouncements on energy policy commonly started with an assertion that increased energy consumption was essential for economic growth. According to this view of the world, for a given country the ratio, commonly termed the energy intensity of the economy, of energy consumption (measured as total primary energy – TPE) to gross domestic product (GDP) was fixed. Consequently, if the economy was to continue to grow, energy demand would also grow, and hence energy supply would have to be expanded to meet the growing demand. This was a belief grounded on the experience of the 1950s and 1960s, when world-wide demand for energy grew each year at a high and generally steady rate. Following the 1973 oil shock, it was predicted that demand would continue to grow and that supply would have to be obtained from alternatives to oil, such as nuclear fission, coal and oil shale. Expectations in Australia for a boom in demand for these energy sources were reinforced by the second oil shock of 1979.

These expectations were not fulfilled; the belief in a fixed TPE:GDP ratio proved to be an unfounded generalisation from the special circumstances of the immediately preceding period. What has actually happened since 1973 in OECD countries is that average rates of economic growth have reduced somewhat, while average rates of growth in energy demand have reduced decisively. For OECD countries as a whole, the energy intensity of their economies, which had been fairly constant during the 1960s, fell by about 25 per cent between 1973 and 1988 (International Energy Agency 1991). In Australia, energy intensity fell by about 13 per cent from a peak in 1977–78 to 1985–86. These percentages represent the amount of energy *not* required to achieve a given level of economic output. The gap between official projections of the mid-1970s and the reality of the late 1980s is even wider. For the OECD as a whole, energy demand in 1985 was only two thirds what had been projected ten years earlier, and in 1990 it was again two thirds what had been projected in 1977. In Australia, actual demand in 1984–85 was 35 per cent lower than had been projected ten years earlier. These are big changes in expectations over a short period, and represent enormous quantities of energy which have not had to be supplied.

This experience has two extremely important implications for current considerations of energy policy in the 1990s and beyond. Firstly, it has meant that the possibility of imminent energy resource exhaustion is no longer an important policy issue, while concerns about pollution at local, regional and global levels have assumed far greater importance. Secondly, it has demonstrated the importance and effectiveness of demand measures, and particularly improvements in the efficiency of energy use, as a means of achieving long-term energy policy objectives.

Improvements in the efficiency of energy use were not the only cause for the reduced energy intensity during the early 1980s. The other important factor was structural change within the economy, meaning a decline in the relative contributions to economic output of such energy-intensive activities as primary metal and cement manufacture, and an increase in the relative contribution of less intensive activities such as information services. For example, the International Energy Agency (1991) quotes the results of a study which concludes that structural shifts accounted for one third of the decline in energy intensity of the US economy from 1972 to 1985, and efficiency improvements for two thirds. Australia differs from other OECD countries in some important respects. A study by the Australian Bureau of Agricultural and Resource Economics (Wilson, Luan and Bowen, 1993) shows that structural shifts actually caused a sharp increase in final energy consumption of about 7% over a short period in the mid-1970s, since when they have caused a gradual decrease of about 6% up to 1990–91. Technical efficiency, by contrast, caused an 11% decrease between 1977–78 and 1985–86, but the trend has reversed since then, with a deterioration in efficiency causing an increase in final energy consumption of nearly 6% up to 1990–91.

In hindsight, the negative trend in energy efficiency during the late 1980s can be seen as a symptom of the general fall in public awareness of energy as a major policy issue over this period, against a backdrop of oil prices falling steadily and, at times, precipitately. However, energy policy has re-emerged in the last few years, largely because, notwithstanding slower rates of growth in energy consumption and significant efforts to reduce the damaging environmental and public health effects associated with the supply and use of energy, it has become apparent that these effects are still extremely serious. They are proving to be both more numerous, more damaging to the global environment and more difficult to control than had previously been thought. Possible global climate change via the enhanced greenhouse effect has played a major role in this. Many environmental effects can be reduced by the use of the appropriate technology; pollutants can be removed from exhaust gas streams or fuels can be processed before use to remove sulfur, trace metals and other contaminants (for example 'clean coal' technology). The cost is typically less than the cost of the environmental damage averted, but this process, which represents an internalisation of at least some of the costs of pollution (making the polluter pay, through energy prices) has forced energy policy makers to look for other, less costly options.

It is in this context, and on the basis of improvements in the efficiency of energy use which have already been achieved, that the focus of energy policy attention has now shifted decisively away from supply-oriented approaches and in favour of emphasis on demand measures. It should not be thought that the improvements in efficiency already achieved have in any way exhausted the potential for further improvements. To give just one example, the Australian Auditor-General (1992) has reported on a Commonwealth of Australia government program, initiated in 1984, to improve the energy efficiency of government buildings, including retrofit work. By 1987, expenditure on retrofits totalling $3.1 million had been committed and many more projects were awaiting approval. However, because of a change in government budget priorities, no more funds were made available. Subsequent evaluation of the work which had been able to be completed, revealed an average payback period on the $3.1 million of 18 months, entirely from savings in energy costs because of efficiency improvements. Clearly, the budget action precluded many more similar savings being realised.

The considerations which favour adoption of demand, or energy conservation, measures, and explain why they have gained increased favour, were well expressed in a report by the International Energy Agency (1987: 7):

> Energy conservation is important for long-term economic well-being and security because:
> - energy conservation will extend the availability of energy resources that are depletable;
> - there is likely to be a return to tightening energy markets before the end of the century; energy conservation will delay and lessen its impact;
> - energy conservation reduces the environmental consequences of energy production and use in a way which is consistent with energy policy objectives;
> - investment in energy conservation at the margin provides a better return than investment in energy supply;
> - investment in energy conservation can often be undertaken in small increments and is therefore flexible at a time when the energy outlook is uncertain.

It is these considerations, notably the last two points, which have been largely responsible for the rapidly growing commitments of the electricity supply industry to demand management activities. Demand management, or demand-side management in US terminology, refers to an array of measures which supply authorities undertake to meet the requirements of their customers by reducing or changing the demand for electricity, rather than increasing the supply. Measures most commonly employed involve reducing demand by increasing efficiency of use; offering rebates on the price of compact fluorescent light bulbs is a familiar example. However, it is important to appreciate that in the electricity industry the term demand management also refers to measures which shift the time at which demand occurs, from peak to off-peak times. Such measures increase the economic efficiency of the supplier, by making better use of power stations and other capital equipment, but do not necessarily use energy more efficiently. Off-peak domestic water heating is the most widespread example of this type of demand management. Nevertheless, it is the demand-reducing/ efficiency-increasing type of demand management which is now receiving most attention from the electricity industry, justified by the cost savings it can deliver in the form of avoided cost, that is, the opportunity to postpone capital investment in new supply capacity.

It is important at this stage to define some of the other terms which are used in discussions of demand-oriented energy policy. In much of the literature on the subject the terms 'energy efficiency' and 'energy conservation' tend to be used interchangeably and/or without definition. More often than not, however, the term energy conservation is seen as having a wider meaning which includes, but goes beyond, energy efficiency improvement. Energy end use efficiency improvement may be usefully defined as reduction from some base level of the energy intensity required by a particular piece of equipment or task, where performance is specified. Energy conservation includes efficiency improvement and also demand reduction, which is defined as reduction in the energy intensity of a given activity by reducing the activity or changing the specifications of the activity, e.g. temperature.

For example, as applied to domestic hot water, energy efficiency improvements could be achieved by upgrading the insulation on the storage tank, or by replacing an older type of gas water heater by a newer so-called high efficiency gas water heater. Demand reduction could be achieved by such measures as fitting a low flow shower head, taking shorter showers, washing clothes at a lower temperature. This distinction between efficiency improvements and demand reduction at the micro-economic level is analogous to the distinction between efficiency improvement and structural change at the level of the macro-economy.

The distinction is most important in the context of analytical studies of energy policy options. The relative efficiencies, for a given output, of two alternative types of energy-using equipment which are intended to deliver the same output, such as two different types of refrigerator, is essentially a technical issue, susceptible to relatively objective specification. Uncertainties about the behaviour of energy users are largely confined to the decision to buy a more efficient model.

By contrast, demand reduction typically involves larger and continuing changes in behaviour, which can be considered to amount to 'lifestyle' changes. It is therefore much more difficult to design policies intended to induce demand reduction, in this sense, without imposing intrusive restraints on individual freedom of choice. On the other hand, societal values and attitudes are continually changing, and it may well be that, as awareness of environmental concerns intensifies, quite significant behavioural changes will occur spontaneously. However, in any study which attempts to project how future energy demand may change or be changed in response to particular policies, behavioural demand reduction will be subject to much greater analytical uncertainty than efficiency improvement. It is because of this analytical difference between efficiency improvement and demand reduction that policy discussions in recent years have increasingly stopped using the term conservation, which normally includes some degree of demand reduction, and replaced it with the more narrowly defined efficiency improvement.

That is not to say that analysis of the scope for efficiency improvement is a simple or non-contentious task. On the contrary, arguments about the extent and cost of improvements in the efficiency of energy use have been probably the single greatest source of debate in the growing body of greenhouse/energy studies done so far in Australia. In effect, this debate is the current expression of the long-standing argument between supply-oriented and demand-oriented approaches to energy policy. The modern supply-siders assert that any measures to increase the efficiency of energy use significantly, and thereby reduce the required supply of energy and the emissions of greenhouse gases associated with supply, will cost more than energy use technologies currently in place. They therefore express scepticism, if not outright denial, that so-called 'no regrets' measures to reduce greenhouse emissions exist (or, for that matter, that these exist for other environmental impacts of energy supply and use). ('No regrets' measures are actions which are economically attractive even if no economic benefit is attributed to emission reduction, that is, measures which should be undertaken irrespective of any concern about the greenhouse effect.) If this cynicism or denial is believed, it follows that any moves to reduce emissions will impose higher costs on individual consumers and the economy as a whole.

Much of the argument in support of this proposition has been advanced at the level of theory, or ideology. It is not appropriate to engage this particular debate in the present chapter. However, it is useful to review the different approaches which have been used to address the issue at the more practical level. At least, three different approaches have been used.

One is to abstract from the issue entirely by modelling at the macro level only. This in essence is the approach taken by all the various studies of different levels of carbon taxes, which proliferated during 1991 and 1992 in the context of the debate over greenhouse response policy. Australian examples include Marks *et al.* (1991), often termed the CRA study, work undertaken by the Industry Commission (1991) using a special version of the ORANI model called ORANI-Greenhouse, the use of ORANI-F by the Ecologically Sustainable Development Working Groups (Brooker and O'Meagher 1991), and a paper put out by the Tasman Institute (Moran and Chisholm 1991). These studies have concluded that imposition of a carbon tax to restrain greenhouse gas emissions would impose quite heavy costs on the economy as a whole. This is because the models used almost invariably achieve emission reductions simply by choking off aggregate demand for energy and with it great swathes of the economy. Detailed examination of models reveals that the technology parameters are inadequately specified to take account of the extent of the technical potential for using energy more efficiently and insufficiently disaggregated to take account of the very great variety of technical possibilities for efficiency improvement in different energy-using sectors. The models are also insufficiently disaggregated to allow for the wide variations in price elasticity of demand for energy between different economic sectors. In other words, the results of the modelling are implicit in the structure of the model, and in no way constitute proof of the non-existence of cost-effective opportunities for efficiency improvement. The same criticism applies to econometric studies, based on historic time series, which claim to show that market imperfections and institutional barriers to energy conservation are insignificant (for example, Ingham *et al.* 1991).

The other two approaches which have been taken in greenhouse energy studies recognise the importance of analysing energy demand at a disaggregated level and trying to include technological characteristics of energy-using processes. The second approach is 'maximum economic potential', which has been used in a number of studies in Australia and internationally. The most comprehensive and detailed Australian study using this approach is the one undertaken by consultants for the Australian Commission for the Future (1991). As the name suggests, this approach   measures the energy which could be saved by efficiency improvements associated with particular technologies in particular sectors or applications, using as the cut-off for savings the test of whether the efficiency technology costs less than purchased energy – that is, additional supply – using a uniform, standardised discount rate, such as 8 per cent. In effect, the approach assumes that all energy users will act as fully rational economic agents and that there are no information or transaction costs associated with choosing efficient energy use rather than additional energy supply. That is, that markets for energy services are perfectly competitive, in the technical economic sense. The advantages of the approach are:

- it is conceptually 'clean';
- if the discount rate is set equal to a social rate (however defined) and it is also assumed, under integrated least cost planning criteria, that the same discount rate is applied to supply-side options, then the calculated level of energy savings is the socially optimal level.

The disadvantages, so far as presenting realistic estimates of costs and benefits is concerned, are that it is based on a highly idealised model of the market for energy services, and in particular:

- it represents a theoretical upper limit (for a given technology), in respect of both the total amount of energy savings and the rate of penetration of efficient technologies;
- it excludes the costs of demand management programs and other measures, which would in most cases be needed to achieve high rates of penetration of efficient technologies, i.e. to overcome the information and transaction costs of energy users.

A number of studies try to get round this problem by adopting a third approach, involving the insertion of the analyst's judgement of the rate and extent of penetration of the new technologies. This approach was used in a study undertaken by consultants for the Business Council of Australia (1991). While possibly more 'realistic' than the maximum economic potential approach, this approach is markedly inferior in terms of methodological clarity. For each sector/technology, a single figure, based on the analyst's judgement, conflates the effects of technical performance with consumer behaviour, the extent of take-up under 'business as usual', possible policies and their effects, etc. As a result, the numbers generated are no more than opinions.

This chapter is concerned with energy conservation in activities other than transport. For the most part, this may be thought of as the energy used in manufacturing processes, in or by buildings for thermal conditioning and lighting and by equipment situated inside buildings, such as household appliances. These categories of energy use encompass what are most commonly grouped as the industrial, commercial and residential sectors. However, non-transport energy use also includes energy used in primary production, in mining and mineral processing, in construction, in electricity and gas generation, transmission and distribution, and in the pumping of water and sewage. The electricity and gas supply options are confined to technologies which can be applied to make existing supply systems more efficient, and also technologies for the simultaneous production of electrical and heat energy on a consumer's premises (cogeneration). Options involving the choice of completely different energy supply technologies, such as solar and wind electricity generation and solar heating, are dealt with in later chapters.

Most of the greenhouse energy studies in Australia, using either of the two disaggregated approaches described above, have looked at the manufacturing, commercial and residential sectors in a more or less comprehensive manner. There are also a few studies which have tried to avoid the defects of both the approaches described – that is, the idealised market assumptions embodied in the maximum economic potential approach and the personal judgements and opinions embodied in the 'realistic' assessment approach. These studies embody separate analyses of the market share captured by a new technology, on the basis of separate payback or rate of return acceptance schedules for different classes of customers over time, and the rate of diffusion of the new technology into the market. By this means, allowance can be made for consumers adopting discount rates which are substantially higher than the social rate, for consumers who prefer old, low-efficiency technology regardless of potential cost savings available from new technology, and for delays in the spread of consumer awareness about new technologies.

This level of modelling sophistication is achieved by the use of specialist proprietary software developed for the analysis of electricity supply industry demand management programs. Such studies have to date been confined to a limited range of electricity use technologies in individual states of Australia.

In the later part of this chapter, the quantitative results of some of these studies using each of the different approaches are summarised. Before presenting the results, however, it is necessary to say something about the type of technologies, and their relationship to current patterns of energy use. Some technologies are applicable across a wide range of energy-using economic activities or sectors, while others have limited, though not necessarily unimportant, application. The number of separate technologies and/or sectors is potentially very large. For example, studies by Southern California Edison, an electric utility in the USA, of CO_2 emission reduction options used a database of 188 separate demand-side technologies (Shin and Sioshansi 1991). Only electric technologies are included in this number.

It is most convenient to group the technologies by major sector of use and then by the type of activity or process to which they are applicable.

Here, the sectors discussed are manufacturing, energy use in or by buildings, and equipment used in buildings. In considering the various technologies in an energy policy context, two characteristics of a technology are most important. One is the relative improvement in energy use efficiency which it can provide; the other is the total amount of energy used nationally by the activity or process to which the

Table 3.1 *Approximate proportionate end use of energy by sectors and activities*

Sector/activity	Activity proportions (%)	Sector totals (%)
Residential		12.2
Space heating	42	
Space cooling	1	
Hot water	26	
Cooking	8	
Refrigerators	6	
Freezers	2	
Other appliances	4	
Lighting	4	
Electronic and other	6	
Commercial		5.5
Space heating	32	
Space cooling	18	
Ventilation	15	
Pumping	3	
Lighting	18	
Hot water	7	
Cooking	3	
Office and other equipment	4	
Lifts, etc.	1	
Manufacturing		33.8
Electrolysis	8	
Electric motors	13	
Smelting	13	
High-temperature firing	15	
Metal processing	4	
Lighting	1	
Low-temperature processes	47	
Agriculture, forestry, fishing		2.3
Mining		4.9
Construction		1.6
Water supply/sewerage		0.2
Transport		39.5

Sources: Consultancy studies commissioned by the Ecologically Sustainable Development Working Groups (see text); Australian Bureau of Agriculture and Resource Economics (1991); author's estimate in some places.
Notes: Based mainly on 1987–88 data; residential data 1986–87. Tools may not sum due to rounding.

technology applies. A technology which can deliver a small relative saving to an activity which uses a very large amount of energy is more important for policy purposes than one which can offer large relative savings to an activity of only minor importance in terms of total energy use.

An indication of the importance of the area of application for each technology in this sense is provided by the data in Table 3.1. Note that this only shows the end use of energy. It does not show energy used in the extraction of energy minerals or, very importantly, energy used and/or lost in the processes of energy conversion, transmission and distribution, that is, in electricity generation, oil refining, coke ovens, etc., and all transmission and distribution by cables, pipelines, and so on.

Because energy use/loss in electricity generation is particularly high (see Table 2.4), an end use which uses a high proportion of electricity implies a much higher primary energy use than an end use which uses mainly gas or petroleum. This makes the residential and commercial sectors, where electricity is the principal energy source, more important in terms of primary energy (and therefore also CO_2 emissions) than might appear from consideration of Table 3.1 alone.

The following summary listing is in no way comprehensive. It is intended to give an indication of the most important types of technical changes which could be used to improve energy use efficiency in each of the main areas of energy use. Much of the information on technologies is taken from a series of studies commissioned by the Ecologically Sustainable Development Energy Use Working Group (EMET Consultants 1991; Energy Policy and Analysis 1991; McLennan Magasanik Associates 1991; George Wilkenfeld and Associates 1991). Some of this material is summarised in the working group's final report (Ecologically Sustainable Development Working Groups 1991).

Manufacturing

In very broad terms, the manufacturing sector may be divided into process industries and fabrication industries. The former use larger quantities of energy to heat or electrolyse materials so that they undergo chemical and/or physical changes into other, desired forms; examples are steel, aluminium, cement, paper, chemicals. The latter – fabrication industries – use more modest quantities of energy, mainly in electrical motors, chiefly to move materials in space; examples are textiles, motor vehicles, electronic goods.

Metal smelting

Smelting is the chemical reduction of metal ores to primary metal by means of chemical reactions at high temperatures using carbon, usually in the form of coke, as the reductant. The production of iron at integrated steelworks is by far the largest energy user in this category, principally because the tonnage of primary steel produced is so large, compared to all other primary metals and to most other materials also. Because of the nature of the equipment used, opportunities to achieve significant efficiency improvements only occur in association with major additions of new plant or replacements of existing plant. Because these are infrequent events, possibilities for short-term efficiency gains are very limited. The main fuel used for smelting is coal, mostly in the form of coke. Some of the technologies are currently judged to be only marginally economic or sub-economic, at current costs of coal. All of the technological changes described here involve or would be associated with major investment programs:

- Dry-quenching of coke, using inert gas instead of quenching with water, would reduce steel-making energy consumption by about 3 per cent, and is already used in Japan.
- Blast furnace top gas recovery turbines extract the energy in the hot, pressurised gas stream which emerges from the top of the blast furnace (currently the gas is collected and burnt for process heating, i.e. the chemical energy is used, but not the thermal/mechanical energy). Overall energy saving would be less than 1 per cent, but in the form of electricity.
- Improved and extended process control, building on the control systems already in place, could achieve savings of a further 2 to 5 per cent over the next 10 to 15 years.
- Direct reduction is an entirely new approach to iron-making which does away with the blast furnace and the preliminary processes of coke production and iron ore sintering. Several different direct reduction processes are under development around the world, including one in Australia (by CRA). Useful, though not enormous reductions in energy use should result, but it is likely to be 20 years before a steel plant using this technology is built in Australia.
- For smelting of copper, lead and zinc, several new processes with significant energy efficiency improvements have been commercialised in recent years, including the Isasmelt process, developed by Mount Isa Mines Ltd from research by CSIRO, which reduces energy required for smelting copper and lead by about 25 per cent. Over the last few years all Australian copper-refining capacity has been converted into Isasmelt and other high efficiency processes. Some lead and zinc smelting also uses Isasmelt, but the majority uses older and intrinsically less efficient processes.

Electrolytic processing

Production of aluminium is by far the most important electrolytic process; others include electro-winning of zinc, electrolytic refining of copper, chloralkali production and metal finishing processes (electroplating and anodising). As with smelting, major improvements in energy efficiency can only be achieved when plants are completely rebuilt; opportunities for incremental efficiency improvements are very limited. This characteristic has important policy implications, because it means that a company, such as an aluminium producer, once established, will be very unresponsive to changes in electricity prices. If prices decrease, it cannot increase production, though it may think about building a new plant if it is confident that prices will stay low over a long period. Conversely, if prices increase, it will have to absorb the extra costs and, if it cannot, shut down and start again, possibly in another country. In other words, the market, as conventionally defined in terms of changing prices, is not an effective instrument for influencing the use of electrical energy by the aluminium industry in the short or medium term. Technological improvements include:

- Drained cathode is a concept for completely redesigning the electrolytic cell so that aluminium metal is drained from the cathode to the bottom of the cell as soon as it is formed. It is estimated that this could reduce energy consumption by up to 15 per cent, but as yet the process requires further development before it can be commercialised.
- New more efficient technologies of various kinds, to replace the Hall–Heroult process which has been used for over a century, are under constant investigation, but none are near to commercialisation.

High-temperature firing

This activity includes a wide variety of production processes involving the heating of non-metallic and ceramic materials to high temperatures, usually above 1000°C. The most important, in terms of energy use, are the calcining of alumina, production of cement, bricks, other clay products and glass, and the sintering and roasting of iron and other metallic ores. A number of energy-saving measures exist:

- In general, a wide variety of measures is available relating to the production of glass, cement, clay products, gypsum, lime, alumina calcining and carbon anode baking, including improved furnaces, more recycling of raw materials and improved process control.
- Significant evergy savings have been achieved in the cement industry

in the last few years with the closure or conversion of almost all plants using the old wet process technology, which typically uses about 75% more energy than the dry process. Further energy savings could be achieved by the greater use of extenders such as fly ash. By international standards, Australian cement companies use little fly ash; its extended use would have additional environmental benefits through reduction in the quantities of fly ash currently dumped in power station ash ponds.

Electric motors

The use of electric motors in an enormous variety of applications probably accounts for over 10 per cent of all energy used in manufacturing (accurate figures are not available) and a much higher proportion of greenhouse gas emissions, given current coal-fired electricity generation technology. Electric motors are also used extensively in other economic sectors, notably mining (including natural gas processing), the commercial sector, water supply and sewerage, and electricity generation. The technologies described below would be applicable in these sectors:

- Matching motor size to load. Many or most motors are oversized relative to their maximum load; since electric motors are more efficient when operated at higher loads, efficiency gains can be made by appropriately matching motor size to load.
- High efficiency motors. Efficient motors achieve lower iron and copper losses, essentially by using more or better quality of each material. There is now only one manufacturer of electric motors in Australia, specialising in the limited market for very large motors. Most motors are therefore imported and importers only carry standard motors in stock; efficient motors must be specially ordered.
- Variable speed drives (VSDs). These are electronic devices which enable the output or speed of motors to be closely adjusted to the required level for applications where load varies widely. They are an alternative to mechanical brakes, baffles, etc., which simply divert unwanted output away from the application and hence waste energy. VSDs offer substantial non-energy benefits to many manufacturing processes, in terms of better process control, longer equipment life, etc. Applications where VSDs are particularly suited include crushing and mixing, conveyors, fans, pumps and compressors.
- Mechanical efficiency in drive systems, etc. Improvements in the efficiency of motor applications, e.g. mechanical drive trains, pumps, and 'good housekeeping' type measures, e.g. shutting off

compressed air and conveyor systems when not in use, eliminating leaks in pipework and compressed air, can reduce electrical energy requirements by reducing motor loads.

High-temperature metal processing

This activity involves the melting of metals in steel and ferro-alloy 'mini-mills' using electric arc furnaces, and in foundries, and the heating of metals for hot working and heat treatment. Total energy is smaller than most other categories of manufacturing activity, and most of it occurs in the steel industry. Thus:

- Near net shape casting and direct linking in steel rolling mills are processes which, in essence, allow hot steel to move directly from the furnace to the rolling mill, without the cooling and reheating required at conventional steel mills. The two are strict alternatives. BHP is currently nearing the end of a 15-year investment program in continuous casting at all its steel mills, which eliminates one stage of reheating. These technologies represent the next major stage in investment.
- Improved furnace design, including use of regenerative or self-recuperative burners, ceramic fibre insulation and installation of advanced control systems are applicable across much of high-temperature metal processing. While some plants currently use these technologies, the majority do not.

Low- and intermediate-temperature process heat

This activity category uses more energy than any other manufacturing activity, amounting to over 10% of total Australian final energy use. Most of the energy is used to heat water to produce hot water and steam, which are then used for an enormous variety of purposes. The food and beverage industries are particularly large users of low-temperature process heat, but virtually every factory in the country uses some low-temperature heat. A number of options can be considered:

- Improved process instrumentation and control, elimination of leaks, increased insulation are all tried and proven approaches for which there remains considerable scope in many factories. Monitoring energy use, normally by installing increased metering, and setting targets for energy use (collectively termed monitoring and targeting) is a management approach of proven effectiveness (Ashton 1990).
- More extensive changes to plant design and operation, involving either retrofits of existing plants or when building a new plant, and

making use of a process integration and analysis technique such as Pinch Technology, can usually achieve larger savings (Ashton 1990).

- The replacement of indirect heating using steam distributed from a central boiler, by gas or electric direct heating technologies, reduces heat losses in the boiler and the distribution system, and may allow the use of technologies which apply heat more precisely to where it is required.

- The combined production of heat and electricity at a manufacturing plant (or a large building), termed cogeneration, is a means of making use of much of the heat energy which is unavoidably lost as waste heat at a conventional thermal power station. Cogeneration is most economic at sites where there is significant consumption of both heat and electricity. It can be based on gas turbine, internal combustion engine or steam turbine, the choice depending on specific circumstances, such as the total heat use and the temperature at which heat is required. The economics are very dependent on the price paid for purchased electricity, which would be displaced by cogenerated electricity.

New technologies

Some new technologies, particularly electrical technologies, have the potential to undertake processes in completely different and much more efficient ways than established technologies, with reductions in total energy use, though often increases in electricity use:

- Membrane separation processes are a way of separating out solid materials dissolved and/or suspended in liquids. The major and most energy intensive method of achieving such separation is distillation, evaporation and drying, which is said to have accounted for 15% of total industrial energy use in the USA in 1985 (International Energy Agency 1991). Membrane separation requires greater use of electrical energy for pumping liquids through membranes, but reduces total energy use.

- Heat pumps, if correctly deployed, can be a useful means of reducing overall energy requirements for many industrial processes by recovering heat energy which would otherwise go to waste (Berntsson 1989). As with membrane separation, the use of heat pumps essentially involves the partial substitution of electricity for a heating fuel, i.e. typically natural gas, plus a significant reduction in total energy requirements for the process or plant concerned.

- Other new electrical technologies offering the potential of overall efficiency improvements involve ways of applying heat with high thermal efficiency. Compared with the non-electrical alternative, which is typically a gas-burning combustion device, they frequently offer non-energy benefits such as less pollution in the work place and improved process control. In terms of energy use, many of these technologies can greatly increase end use efficiency. Examples include induction heating, eddy current heating, microwave heating, infrared heating and ultraviolet curing.

Materials recycling

Increased recycling of materials such as steel, aluminium, paper and glass would, in general terms, reduce energy requirements, since less energy is required to recycle than to produce virgin material. Australia is currently a net exporter of scrap steel, aluminium and newsprint (Industry Commission 1990). The potential implications for use of electricity of greater use of scrap within Australia are complex and need to be considered on a case by case basis:

- *Steel and aluminium.* Australia is a relatively low cost producer and net exporter of these metals; it is therefore logical to assume that any further recycling of steel, through electric arc steel-making, or aluminium will be additional to, not a replacement for, conventional production capacity.
- *Paper.* There is appreciable scope for greater recycling of all kinds of paper in Australia and paper manufacturers have proposals in place to significantly increase the proportion of recycling and reuse. This would reduce consumption of electricity, particularly at mechanical pulp mills used to make newsprint, and heating fuels, particularly at chemical pulp mills used to produce packaging and writing papers.

Energy use in or by buildings

Included in this sector is energy used for heating, cooling and lighting buildings. All of the technologies mentioned here have been proven in commercial operation – they are not new or experimental. However, most of them are not widely used in the existing building stock, and retrofitting in association with building refurbishment offers substantial opportunities for energy savings. Many of the technologies are not even generally installed in new buildings, although the great majority would be cost effective in most circumstances.

Heating, ventilation and cooling (HVAC)

HVAC systems in commercial and community service buildings offer a range of energy saving potentials:

- The use of more efficient cooling equipment in air-conditioning installations.
- Use of high efficiency (outside air) cycles, which use outside air to cool the internal building space when ambient temperatures are suitable.
- Wider use of variable air volume (VAV) systems, which control the temperature by varying the volume of air supplied, rather than the temperature of the air (which typically involves cooling and reheating, to avoid condensation problems with excessively humid air). However, the major energy savings arise from the reduced ventilation load.
- Substitution of heat pumps for resistance heating in smaller buildings (larger buildings generally already use heat pumps or gas heating).
- Recalibration and regular maintenance of controls.
- Installation of optimum start/stop controls, so that HVAC systems do not operate constant hours throughout the year, but instead operate fewer hours during mid-season periods.
- Reflective window foil in existing buildings to reduce summer heat gain through windows, possibly in combination with external shading, weatherproofing, etc.
- Retrofitting of insulation in community service buildings, many of which are poorly insulated.

Lighting in commercial and community service buildings

Some of these measures, such as the use of more efficient ballasts in fluorescent lights, are also applicable to lighting in factories, i.e. energy use which would normally be classified under manufacturing. Major areas are:

- Replace incandescent lights with fluorescent (including compact fluorescent) lamps. Although standard fluorescent light fittings account for the greater proportion of lighting in the commercial and services sector, there are still a number of incandescent light fittings. As a broad generalisation, most of these could be replaced by compact fluorescent bulbs, or by a mixture of standard fluorescent fittings, compact fluorescents and fluorescents with dimmer controls.
- Replacement of standard ferromagnetic ballasts on fluorescent lights with low loss ferromagnetic ballasts and the addition of reflectors or high efficiency luminaries to standard fluorescent light fittings. These are two quite separate measures, but undertaking them together greatly economises on the labour costs of retrofitting programs.

- Installation of controls to allow office lighting to be controlled on the basis of several zones per floor (termed zone control).
- Installation of sensors and controls to allow perimeter lighting to be switched off when daylight is sufficient (termed daylighting control).
- Installation of switches, etc., to control lighting levels for individual rooms, offices, etc.
- Reduction of excess lamping in areas of office buildings used as corridors, etc.
- Installation of controls to eliminate unnecessary use of lighting of external and common areas when a building is vacant.
- Regular maintenance of lighting systems and reflective surfaces to ensure that luminance is maintained at high levels and allows the number of lamps to be reduced.

Residential space heating and cooling

Major areas for energy saving include:

- Incorporation of well known thermal efficiency design features, including correct building orientation and shading, correct placement of glass, appropriate thermal mass and ceiling and wall insulation in all new houses. The optimal amount of each of these measures will depend on the climate in the location where the house is built. Some of these measures, such as ceiling insulation and window shading, can normally be economically retrofitted to existing houses.
- Installation of high efficiency air conditioners. Both room and central air conditioners can be made more efficient by increasing condenser and evaporator areas and the use of more efficient fan motors and compressors.
- Fuel/technology substitution for space heating, that is, replacing electric resistance heating by heat pumps, gas or solid fuels. All these changes significantly increase the thermal efficiency of space heating, but are not always clearly economical when compared to off-peak prices for electricity available in most parts of Australia. The use of solid fuels in urban areas has the additional problem of causing significant air pollution, even when used in the best available types of solid fuels heaters. The overall energy savings through fuel substitution will, of course, depend on the total requirement for heating energy. For this reason, these measures are likely to be more effective as retrofits in older existing houses than in new houses which incorporate energy-efficient features to reduce the requirement for all forms of heating.

Residential lighting

This includes:

• Replacement of the most intensively used incandescent lamps in houses with compact fluorescent lamps. Although efficiency gains can be made by replacing all lamps, the much higher cost of compact fluorescents means that economics strongly favours only replacing the most used lamps, which will of course deliver the greatest energy savings.

Equipment used in buildings

Although a great variety of equipment is used in commercial and residential buildings, relatively few types account for the majority of energy use. In residential buildings, water heating is by far the most important, followed by cooking and refrigeration. In commercial buildings, water heating is normally much less important, but in some types of service buildings such as hospitals and hotels/motels, it is just as important as in residential buildings. Office equipment is relatively more important in commercial buildings, and refrigeration is very important in buildings used for selling or supplying food or drinks.

Equipment in commercial buildings

In commercial sector refrigeration, systematic and cyclic maintenance can be used to achieve higher levels of operating efficiency.

With respect to office equipment, it should be noted that quite large increases in efficiency of individual pieces of equipment are being realised as new equipment comes onto the market, such as the use of liquid crystal technology instead of cathode ray tubes in VDUs. However, this trend is offset by the very rapid growth in the total amount of equipment in use. Efficiency gains can be made through the installation of control systems to turn off office equipment when not in use over prolonged periods.

In commercial cooking, improved management of kitchens can reduce energy waste, and the increased use of microwave ovens in place of gas and electrical resistance heating in stoves and ovens offers further savings. Energy use in commercial water heating can be reduced through the installation of heat reclaim systems, which allow hot water to be obtained from the waste heat expelled by air conditioning chiller units.

Fuel/technology substitution, including use of heat pump water heaters or gas water heaters instead of resistance heating, can also be used, as can the installation of flat plate solar collectors to provide pre-heating. Existing hot water storage tanks can be replaced with new tanks by using improved construction materials and more insulation.

Equipment in residential buildings

A large number of technical options are available to decrease the energy used for residential water heating, and these fall into two general categories. The first involve more efficient use of water and the reduction of losses associated with storage water heaters. Major options are the use of additional insulation blankets on water storage tanks, and the use of low flow showerheads to reduce hot water consumption.

However, most efficiency gains involve fuel and technology substitution away from electric off-peak and continuous storage water heaters, which in overall terms are the least thermally efficient types. Because there are so many substitution possibilities and combinations, quantitative analysis of the total energy savings potential, and of the associated economics, is very complex. The main technologies available for substitution, other than electric storage, include the following:

- standard gas;
- high efficiency gas;
- electric heat pump;
- solar-boosted heat pump;
- electric-boosted flat plate solar (with either continuous or off-peak boost);
- gas-boosted flat plate solar.

To save energy in residential cooking, more efficient ovens and microwaves incorporating forced convection (fans), increased insulation and improved gaskets around doors are available. Also available are more efficient electric stoves (cooktops) incorporating reflective pans under elements and reduced electrical resistance in element contacts, and new types of electric stove elements, including halogen lamp and induction elements, which may be more efficient overall than conventional (resistance) electric elements or gas elements.

More efficient refrigerators and freezers can be used in households, incorporating improvements in the amount and type of insulation, compressor efficiency and type of control system, and the optimal integration of all components. The most efficient refrigerators currently manufactured in the world, embodying these types of features, are roughly three times as efficient as standard Australian refrigerators (Norgard 1989).

With dishwashers, the main opportunities for efficiency gains come from design improvements which allow the quantity of hot water required to be reduced, including improvements in pump efficiency. With washing machines, the main energy savings opportunities come from reducing average water temperature and greater use of front-loading rather than top-loading designs. More efficient clothes dryers use a moisture termination sensor control, improved insulation and exhaust heat recycling. Finally, efficiency improvement trends associated with technological change in household electronic equipment, particularly television receivers, are expected to progressively reduce the energy used by these appliances.

Conclusion

The most detailed study of the potential energy consumption savings, assessed on a 'maximum economic potential' basis, is that compiled by the Ecologically Sustainable Development Working Groups (1991), on the basis of a series of studies by consultants (referred to previously) of the technical potential improvements in energy use efficiency in selected energy services in the residential, commercial and manufacturing sectors. The studies examined the present populations of energy-using equipment relevant to the respective processes, made use of specified 'frozen efficiency' rates of growth in energy demand (implying specific rates of growth in the populations of energy-using equipment), assessed the energy savings available from the use of selected technologies for each service and estimated the associated costs and benefits. The energy services assessed were as follows:

- Residential: Hot water
                    Refrigerators and freezers
                    Washing machines and dish washers
- Commercial: Heating, ventilation and air conditioning
                    Lighting
                    Other services (hot water, cooking, office and
                         other electrical equipment)
- Manufacturing: Metal smelting
                    Electrolytic processing
                    High-temperature firing
                    Metal melting and other high-temperature metal
                         processing
                    Electric motors and drives

For each of these services or groups of services, a wide variety of technical options were assessed, including, as relevant, efficiency improvements, process changes and fuel substitution.

The economic costs and benefits of the various options were assessed, exclusive of any environmental benefit of energy saving itself. Thus benefits included reduced energy consumption, measured in terms of a schedule of projected resource·costs for electricity, gas, petroleum products and coal, and in some cases associated savings in maintenance costs, increased productivity of capital equipment, etc. Costs were the additional capital costs of the new equipment. The studies were confined to measures which either showed a net benefit or a relatively modest net cost.

These studies used the 'maximum economic potential' approach, and hence in most cases assumed very high rates of penetration of optimally efficient technologies in purchases of new plant and equipment, while excluding administration or incentive costs for programs which might be needed to stimulate the economically optimal choice of equipment. For this reason, the estimates of net benefits/costs of emission abatement could be interpreted as understating the costs. On the other hand, in a number of the studies, notably those dealing with residential sector energy services, the assumed resource cost for electricity was substantially less than the true cost at the customer's meter, so that benefits of energy savings are understated.

It is not possible to specify the estimated total energy savings potential in an accurate or simple way, since the study results are presented in terms of reductions in carbon dioxide emissions. The total carbon dioxide emission reduction potential from all of the energy services studied was about 32% relative to 'frozen efficiency' demand projections, and assuming no change in the present mix of supply technologies. On this basis, carbon dioxide emissions associated with the services studied in 2005 would still be about 11% higher than 1988 emissions. This compares to the Australian Government's interim target of a 20% reduction on 1988 levels by 2005.

A broadly similar conclusion was reached by a study sponsored by the Australian Commission for the Future (1991). This study used the same 'maximum economic potential' approach. It covered all sectors of energy use, but used much more generalised and less detailed estimates of the costs and savings potentials of individual technologies and sectors.

A critique of the ESD studies has claimed that they overstate the potential and understate the cost of achieving such levels of emission reduction by demand-side measures (ACIL Australia 1992). The critique points to imprecision and confusion in the definition of the 'business as usual' case against which savings are measured and argues that the estimates of potential savings exaggerate what is achievable. In its report, ACIL provides an estimate of total carbon dioxide emission reduction potential from the 11 sectors included in the Ecologically Sustainable Development studies which is about 37% lower than that derived from the studies. This lower total is obtained by applying ACIL's own, essentially arbitrary, lower levels of efficiency gain and lower rates of new technology penetration.

In essence, this criticism is simply a restatement of arguments about the existence, nature and causes of the 'efficiency gap' (Grubb 1990). As such, the criticism misunderstands the purpose of the studies, which were concerned with the technical potential for savings, not the savings achievable under the prevailing economic and policy environment. The savings identified were those assessed as being cost effective at a discount rate of 8 per cent, which is generally accepted as an appropriate rate for determining social costs and benefits, but is considerably lower than the discount rate commonly used for private and business decisions about purchases of energy-using equipment.

To achieve the technical potential for energy efficiency improvements will require quite large and rapid changes in purchase decisions by energy users and will in turn require the implementation of a variety of new policies and programs by governments, energy utilities and other parties. These measures will not be costless. They will, however, deliver significant benefits in the form of higher overall efficiency of the Australian economy and reductions in environmental impacts unavoidably associated with energy supply and use.

References

ACIL Australia. 1992. An assessment of the achievability of an Australian commitment to stabilise energy-related CO_2 emissions. In: *Two studies pertinent to Australia's decision on the terms of participation in a global convention on climate change.* Canberra: ACIL Australia.

Ashton, G. 1990. *Energy management in the process industries: project report.* Sydney: Warren Centre for Advanced Engineering, University of Sydney.

Australian Auditor-General. 1992. *Energy management of Commonwealth buildings: Audit Report No. 47, 1991–92.* Canberra: Australian Government Publishing Service.

Australian Bureau of Agricultural and Resource Economics. 1991. *Projections of energy demand and supply Australia, 1990–91 to 2004–05.* Canberra: Australian Government Publishing Service.

Australian Commission for the Future. 1991. *Energy futures.* Melbourne: Commission for the Future.

Berntsson, T. 1989. Heat pumps. In: Johansson, T.B., Bodlund, B. and Williams, R.H. (eds), *Electricity: efficient end-use and new generation technologies, and their planning implications,* pp. 173–216. Lund: Lund University Press.

Brooker, R. and O'Meagher, B. 1991. Economic modelling, ecologically sustainable development and the greenhouse effect: economic aspects of reducing greenhouse gas emissions. In: *Economic modelling.* Canberra: Ecologically Sustainable Development Working Groups.

Business Council of Australia. 1991. *Energy prospects: realistic prospects for improved energy efficiency and the use of renewable energy sources.* Melbourne: Business Council of Australia.

Ecologically Sustainable Development Working Groups. 1991. *Final report: energy use.* Canberra: Australian Government Publishing Service.

EMET Consultants Pty Ltd. 1991a. *Evaluation of costs of options for greenhouse gas reduction in energy use: commercial sector lighting.* Unpublished, available from Department of Primary Industries and Energy, Canberra.

EMET Consultants Pty Ltd. 1991b. *Evaluation of costs of options for greenhouse gas reduction in energy use: commercial sector heating and cooling.* Unpublished, available from Department of Primary Industries and Energy, Canberra.

EMET Consultants Pty Ltd. 1991c. *Evaluation of costs of options for greenhouse gas reduction in energy use: commercial sector process heat and other uses.* Unpublished, available from Department of Primary Industries and Energy, Canberra.

Energy Policy and Analysis Pty Ltd. 1991a. *ESD energy use consultancy: industrial electric motors and drives.* Unpublished, available from Department of Primary Industries and Energy, Canberra.

Energy Policy and Analysis Pty Ltd. 1991b. *ESD energy use consultancy: metal processing.* Unpublished, available from Department of Primary Industries and Energy, Canberra.

George Wilkenfield and Associates. 1991a. *Costs of meeting carbon dioxide emissions targets in residential water heating.* Unpublished, available from Department of Primary Industries and Energy, Canberra.

George Wilkenfeld and Associates. 1991b. *Costs of meeting carbon dioxide emissions targets in residential refrigeration.* Unpublished, available from Department of Primary Industries and Energy, Canberra.

George Wilkenfeld and Associates. 1991c. *Costs of meeting carbon dioxide emissions targets in major household appliances.* Unpublished, available from Department of Primary Industries and Energy, Canberra.

Grubb, M. 1990. *Energy policies and the greenhouse effect. Volume one: policy appraisal.* Aldershot: Dartmouth Publishing Company.

Industry Commission. 1990. Recycling. *Volume I: recycling in Australia.* Canberra: Australian Government Publishing Service.

Industry Commission. 1991. *Costs and benefits of reducing greenhouse gas emissions.* Volumes II and III. Canberra: Australian Government Publishing Service.

Ingham, A., Maw, J. and Ulph, A. 1991. Testing for barriers to energy conservation: an application of a vintage model. *Energy Journal.* 12(4): 41–64.

International Energy Agency. 1987. *Energy conservation in IEA countries.* Paris: International Energy Agency.

International Energy Agency. 1991. *Energy efficiency and the environment.* Paris: International Energy Agency.

Marks, R.E., Swan, P.L., McLennan, P., Schodde, R., Dixon, P.B. and Johnson, D.T. 1991. The costs of Australian carbon dioxide abatement. *Energy Journal.* 12(2): 135–152.

McLennan Magasanik Associates. 1991a. *Unit operation – electrolytic processing.* Unpublished, available from Department of Primary Industries and Energy, Canberra.

McLennan Magasanik Associates. 1991b. *Unit operation – smelting.* Unpublished, available from Department of Primary Industries and Energy, Canberra.

McLennan Magasanik Associates. 1991c. *Unit operation – high temperature firing.* Unpublished, available from Department of Primary Industries and Energy, Canberra.

Moran, A. and Chisholm, A. 1991. *Greenhouse gas abatement: its costs and practicalities.* Melbourne: Tasman Institute.

Norgard, J.S. 1989. Low electricity appliances – Options for the future. In: Johannson, T.B., Bodlund, B. and Williams, R.H. (eds), *Electricity: efficient end-use and new generation technologies, and their planning implications,* pp. 125–172. Lund: Lund University Press.

Shin, J.S. and Sioshansi, F. 1991. *Control strategy to reduce greenhouse gas emissions: methodology development and application for the electric power industry.* Paper delivered at the 14th Annual International Conference of the International Association for Energy Economics, Honolulu, Hawaii, 8–10 July, 1991.

Wilson, B., Luan, H.T. and Bowen, B. 1993. *Energy efficiency trends in Australia.* Canberra: Australian Bureau of Agricultural and Resource Economics Research Report 93.11.

CHAPTER 4

Energy conservation in transport and urban settlements

PETER NEWMAN

Energy conservation to most people is turning off lights and to a limited extent using less heating. Occasionally it might mean having a more fuel-efficient automobile (car). On a thermodynamic assessment of energy it takes an order of magnitude more energy to heat than to light something, and a further order of magnitude more energy to move than to heat. The strategic importance of transport technology and the large potential for energy conservation in this sector is, however, not always so obvious.

This chapter will examine the large variation that exists in the use of transport fuel around the world and how this gives us some clues as to how much potential there is for conservation. It will concentrate on the urban situation where most of the fuel is used (over 50% in Australia) but will look at some intercity options as well. It will attempt to examine the potential for improvement in vehicles as well as the changes that can reduce the need for travel and particularly the need for automobile travel. Policies will be outlined based on a comparison of data from 31 global cities as well as the Australian government's Ecologically Sustainable Development (ESD) process. A particular focus is the final report of the ESD Transport Working Group (1991), and the responses to this from the conservation movement in the form of the Australian Conservation Foundation–World Wide Fund for Nature (ACF–WWF) response publication (Hare 1991).

Why reduce transport energy?

There are obvious direct benefits if transport energy can be saved, such as:

- lessening the vulnerability of a nation to oil supply disruptions, thus improving its sustainability in energy terms and contributing to a more peaceful world that is becoming more and more dependent on Middle Eastern oil (see Campbell 1991, who suggests global oil production will peak around the year 2000);
- minimising the effect of transport-related inflation and the national balance of payments due to imported oil and excessive, wasteful travel (see Vintila *et al.* 1992); and
- reducing the quantity of emissions including those that contribute to the greenhouse effect and to smog.

There are many other indirect benefits that can be derived if the energy conservation is through reduced automobile usage and dependence, such as:

- lowering the noise and hostility of traffic in cities and the huge economic and social costs from road accidents;
- improving the balance between public transport (transit) and private transport (automobiles) thus reducing the public transport deficit;
- improving public health through the greater use of bicycling and walking as transport modes;
- improving the level of accessibility to the transport disadvantaged, i.e. the elderly, children, poor people and handicapped who cannot use an automobile, as well as those in poorly serviced outer suburbs built around automobile use; and
- enhancing the public spaces of a city through a reduced emphasis on private automobiles.

Perspective on urban transport energy use

A perspective on how cities use transport energy is set out in Table 4.1 based on a comparison of 31 cities around the world.

The data show that United States (US) cities use on average 66.6 GJ of fuel per capita from transport compared to 36.3 GJ per capita in Australian cities, 14.7 GJ in European cities and 11.6 GJ in Asian cities. This is an enormous variation, much greater than could be explained by obvious economic factors. The different factors behind the variation are now explored.

Fuel types

The breakdown by fuel shows that gasoline is by far the biggest contributor to transport energy use, but this is most marked in US and Australian cities where the automobile is more dominant; where cities

Table 4.1 *Per capita transport energy use in global cities, 1980*

City	Motor vehicles: Gasoline (MJ)	Diesel (MJ)	% of Total	Public transport: Diesel (MJ)	Electricity (MJ)	% of Total	Total (MJ)
Houston	74,510	9398	100	289	0	<1	84,197
Phoenix	69,907	8842	100	181	0	<1	78,931
Detroit	65,980	2990	99	377	0	1	69,346
Denver	63,465	13,736	99	542	0	1	77,743
Los Angeles	58,474	6396	99	645	0	1	65,515
San Francisco	55,365	9777	98	839	235	2	66,216
Boston	54,182	10,039	99	364	90	1	64,674
Washington	51,242	2122	98	786	194	2	54,343
Chicago	48,246	3742	97	1213	180	3	53,381
New York	44,033	6170	96	547	1340	4	52,090
US Average	**58,540**	**7321**	**99**	**578**	**204**	**1**	**66,643**
Perth	32,610	6822	105	816	0	2	40,248
Brisbane	30,654	5795	98	799	56	2	37,304
Melbourne	29,104	4542	98	247	350	2	34,243
Adelaide	28,790	5402	97	1007	7	3	35,206
Sydney	27,986	5692	97	660	320	3	34,657
Australia Average	**29,829**	**5651**	**99**	**706**	**147**	**2**	**36,333**
Toronto	34,813	12,173	97	979	509	3	48,474
Hamburg	16,671	5942	98	149	313	2	23,074
Frankfurt	16,093	5652	96	216	619	4	22,580
Zurich	15,709	2092	93	160	1080	7	19,041
Stockholm	15,574	5562	93	934	683	7	22,754
Brussels	14,744	6710	94	574	764	6	22,793
Paris	14,091	4856	96	302	485	4	19,733
London	12,426	5171	94	644	480	6	18,721
Munich	12,372	4348	95	249	558	5	17,526
West Berlin	11,331	4050	93	688	396	7	16,465
Copenhagen	11,105	4622	92	843	462	8	17,032
Vienna	10,074	1191	93	347	526	7	12,138
Amsterdam	6842	2941	93	588	101	7	10,472
Europe Average	**13,280**	**3430**	**93**	**543**	**408**	**7**	**14,726**
Tokyo	8488	4583	94	252	723	6	15,346
Singapore	5958	3498	86	1579	0	14	11,036
Hong Kong	1987	2564	84	792	65	16	5409
Asia Average	**5493**	**3519**	**89**	**792**	**299**	**11**	**11,629**

Source: Newman and Kenworthy (1989).
Note: LPG is added to gasoline and is very small.

become more public transport oriented, diesel and electricity become much more significant. Tokyo has 52% gasoline, 33% diesel and 15% electricity, whereas Houston has 89% gasoline, 11% diesel and no electricity.

The breakdown between private and public transport shows an overwhelming proportion of transport energy is from private usage. Public transport uses only 1% of transport energy in US cities, 2% in Australian cities, 7% in European cities and 11% in Asian cities. Diesel has a remarkably uniform pattern across the cities; for example, Hamburg and New York have almost the same per capita motor vehicle diesel use indicating the similar dependence that most cities now have on the light van and truck for urban freight movement.

The major difference between the cities is in the comparative use of gasoline and electricity. Gasoline-oriented cities are heavy energy users whilst cities with any significant level of electricity use in their transport system are low energy users overall. Despite coal-based electricity being less fuel efficient than gasoline (and four times worse than gasoline in terms of CO_2 produced per MJ of transport energy), it does not mean cities with electric transport are worse in energy use or greenhouse gas emission terms; in fact, the reverse is the case. This is because of the nature of the technology and the effect of either the automobile or the train/tram on the city. This difference is fundamental to the concepts being presented in this chapter. It is an important factor in the energy and greenhouse debate where coal is considered to be so much more damaging; if coal is used to provide an electric train or tram system then the city will overall use less fuel and produce lower greenhouse gases. The mechanism for this appears to be land use changes, greater walking and cycling and a combination of trips when using transit: this will be examined in more detail below.

As well, the advantages of electric-based public transport become far greater in terms of greenhouse gas emissions, and other problems like smog and acid rain, when the source of power is renewable fuel. This will become a more significant option in the decades ahead as oil prices continue to rise. The renewable future will be an electric one linking together the many dispersed ways of producing power from wind, sun, plantations, garbage and waves (see following chapters). An electric future will favour electric transit (perhaps supplemented by electric automobiles) as a renewable electricity-based system implies that energy conservation will be taken seriously and this will mean a substantial move towards transit-based cities. This concept will be expanded by reference to the many potential factors that can be used to explain the above variations in fuel use.

Factors explaining urban transport energy use

The various factors that are considered to be related to the use of transport fuel are set out below in relation to the data on urban transport energy in Table 4.1. The emphasis is on explaining the huge variation in use of automobiles between the world's cities.

Vehicle efficiency

Obviously the efficiency of vehicles must impact on fuel use, but how important is it in explaining the variation between fuel use? Table 4.2 adjusts gasoline use for vehicle efficiency; that is, Table 4.2 shows what the situation would be like if all cities had vehicles like US cities. Overall, the variation from US cities to Asian cities in their gasoline use is reduced from a factor of ten to a factor of eight, and US to European cities reduces from four times to three and a half times. Thus this technological factor is relevant, but clearly is not the dominant factor often described in energy conservation literature (see, for example, La Belle and Moses 1982; Chandler 1985). Other economic and planning factors need to be considered.

One of the anomalies in Table 4.1 is the position of Toronto, which has high transit usage but also relatively high gasoline usage – perhaps vehicle efficiency can explain this. When vehicle efficiency is removed

Table 4.2 *Adjusted average gasoline use per capita in cities by region to account for vehicle efficiency, 1980*

Average by cities	Unadjusted gasoline use per capita (MJ)	Average vehicle efficiency (national values) (l/100 km)	Adjusted for average speed in cities (l/100 km)	Gasoline use per capita with US vehicle efficiency (MJ)	
				National values	Adjusted for average speed
Average, 10 US cities	58,541	15.35	19.33	58,541	58,541
Average, 5 Australian cities	29,829	12.50	15.33	33,446	37,612
Toronto	34,813	16.30	21.72	32,784	30,982
Average, 12 European cities	13,280	10.66	16.38	19,123	15,727
Average, 3 Asian cities	5493	7.63	15.05	11,051	7248

Notes: (1) Detailed data on vehicle efficiencies are contained in Newman and Kenworthy (1989).
(2) Adjustments for average speed are made by using $y = 1.0174x + 37.4291$ where y = fuel consumption in ml/km and x is the inverse of average speed in s/km (Kenworthy, Newman and Lyons 1983) and national fuel efficiencies are assumed to be at an average speed of 60 km/h.

as a factor, Toronto's gasoline use is less than Australian cities, consistent with its picture as a city with less automobile usage and more transit than Australian cities (see below). Thus vehicle technology can help explain some of the variations.

It was a feature of the ESD Transport report outlined below that the vehicle technology approach was examined but did not dominate the analysis as in most previous Australian government approaches to this subject. This is a positive sign, as concentrating on vehicle technology means that the major structural questions associated with transport and energy are easily neglected. This was also apparent in a recent Commonwealth Government report, which sets out a range of options for energy conservation (Australian Department of Primary Industries and Energy 1991).

Price and income

The economics of transport is dominated by considerations of price (especially gasoline prices) and income; these are considered to be the major determinants of travel demand, and particularly in determining the level of automobile ownership and through this the level of use. Obviously the price of gasoline and how much spare income you have will be a big factor in determining how people travel. However, the extent of these economic factors is questioned by the data provided in Table 4.3 which makes adjustments for price and income. The close interrelationship between price and income on the size of automobile purchased means that this factor was also included in the calculation.

Table 4.3 sets out expected gasoline consumption if US incomes, gasoline prices and vehicle efficiencies were found in all the 31 cities. If these economic factors were the sole or primary factors then all the cities should have the same gasoline use; they clearly do not. On average, the economic factors explain at the most around half the gasoline use. It can be argued that these long-term elasticities are overestimates as they incorporate some degree of anticipated urban form change, but even they do not adequately explain the variations in gasoline use in the sample. What is suggested by these results is that a purely economic approach to transport matters will be inadequate, and that matters of urban form and provision for the automobile have direct and independent influence on transport patterns. Thus planners who provide the transport infrastructure or who set out the physical plan of a city are *directly and actively* influencing transport patterns; they are not simply reacting to economic factors.

The same conclusion can be found by considering automobile ownership. Table 4.4 shows how this varies, indicating higher values in

Table 4.3 *Average value for per capita gasoline use in cities by region, 1980*

Average by cities	Actual gasoline use (MJ per capita)	Adjusted gasoline use for US gasoline prices, incomes and vehicle efficiency (MJ per capita):		Difference (%) between US gasoline use and adjusted gasoline use by other cities:	
		Short-term elasticities	Long-term elasticities	Short-term elasticities	Long-term elasticities
Average, 10 US cities	58,541	58,541	58,541	—	—
Average, 5 Australian cities	29,829	38,488	43,680	51	25
Toronto	34,813	29,995	26,090	49	55
Average, 12 European cities	13,280	17,082	31,080	71	47
Average, 3 Asian cities	5493	7676	12,340	87	79
Average for non-US cities	17,133	21,450	31,160	63	47

Sources: Pindyck (1979), Dahl (1982), Archibald and Gillingham (1981) and Wheaton (1982).

Notes:

(1) Gasoline consumption elasticities used were:

	Short-term	Long-term
gasoline price	−0.20	−1.0
incomes	+0.11	+0.6

(2) As gasoline consumption elasticities include a component due to vehicle efficiency, it is necessary to subtract this when adjusting other cities for US vehicle efficiencies, otherwise it would be accounted for twice. Vehicle efficiency elasticities used were:

	Short-term	Long-term
gasoline price	+0.11	+1.0
incomes	−0.11	−1.0

Vehicle efficiencies used are national values adjusted for average speed in each city.

In all cases vehicle efficiencies in the long term became more than equivalent to US levels and hence the efficiency factor in the long term is cancelled out.

US and Australian cities than in European and Asian cities; but when automobile usage per capita is compared there is still a large variation. Obviously there are many people who do not need to use an automobile or do not need to use one as much, in European and Asian cities compared to US and Australian cities.

Table 4.4 *Car ownership and usage and their relation to transit in global cities, 1980*

Average by cities	Car ownership per 1000 people	Car usage pass. km/per.	Transit usage pass. km/per.	Total travel pass. km/per.
Average, 10 US cities	533	12,507	522	13,029
Average, 5 Australian cities	453	10,680	856	11,536
Toronto	463	9,850	1976	11,826
Average, 12 European cities	328	5595	1791	7386
Average, 3 Asian cities	88	1799	3059	4858

Source: Newman and Kenworthy (1989).

Having asserted the importance of physical planning it should also be clearly stated that the economic parameters are significant. Kirwan (1992) has re-analysed our data in terms of multivariate regressions and concluded that the price of fuel is the most significant variable in influencing travel patterns. He does, however, admit that there are strong factors influencing travel patterns which are part of the structure or physical form of the city. Obviously pricing does influence travel and needs to be considered in any policy package designed to save transport energy.

The relative provision for transport modes

1. Modal splits

Table 4.5 sets out the variations in transport patterns and in particular their relation to transit provision. The most highly automobile-oriented US cities have virtually no public transport as a percentage of their total passenger kilometres (km) of travel – for example, Houston 0.8%, Phoenix 0.5%, Detroit 0.8%. Even in a city like Denver, where there is a strong policy to encourage bus usage due to the smog, only 1.8% of total passenger travel is by public transport. It is only in the US cities with rail systems that any significant proportion of transport is by non-automobile modes: such as San Francisco 7%, Chicago 8%, New York 14% (the proportion of total transit passenger kilometres by trains is San Francisco 34%, Chicago 67% and New York 78%). At the same time the bicycling/walking proportion for journey to work trips rises (up to 10% in Boston, 6 to 8% in others).

Table 4.5 *Relative performance and provision of transport modes in global cities, 1980*

Average by cities	Transit (of total vehicle pass. km) (%)	Transit service (for journey to work) (%)	Transit (veh. km of service per person)	Road provision (m per person)	Car parking (per 1000 CBD workers)	Average speed of travel (km/h): Car	Train	Bus
Average, 10 US cities	4.4	5.3	30	6.6	380	43	42	20
Average 5 Australian cities	7.5	5.2	56	8.7	327	44	39	21
Toronto	16.7	5.8	81	2.7	198	–	34	20
Average, 12 European cities	24.8	21.3	79	2.1	211	33	43	20
Average, 3 Asian cities	64.1	25.1	103	1.0	67	24	36	15

Source: Newman and Kenworthy 1989.

Australian cities overall are a little less automobile-oriented, though Perth (5% transit, 4% bicycle/walking) is virtually an average US city. Sydney with 14% public transport use is the most non-automobile oriented Australian city with once again a high proportion on rail (70% of transit). Toronto is significantly different to its North American neighbours with 17% transit use (in particular the comparison with its nearest neighbour Detroit at 0.8% is quite stunning).

European cities on average have 25% public transport use for the total passenger transport task (in passenger kilometres) and, for work journeys, 21% of trips are by bicycling/walking. This ranges from 17% public transport in Hamburg to 32% in West Berlin and 30% in Paris and Vienna; for bicycling/walking to work Copenhagen at 32% and 28% in Amsterdam are the highest. In the European cities 55% of public transport passenger kilometres are on trains. On average, people in US cities travel nearly 7000 kilometres further by automobile and 1200 km less by public transport than in European cities. Among other things, this suggests urban travel distances are shorter in Europe and in fact work journey average distances are 30 to 40% shorter in European cities which have an average of 8 km compared with US cities of 13 km and Australian cities 12 km.

All these comparisons are even more striking when the Asian cities are examined where 64% of the transport task is by public transport and 25% of people go to work by walking or biking (35% in Hong Kong). In the modern metropolis of Tokyo only 16% of the people use an automobile to go to work and in the public transport system 95% of passenger kilometres are by train.

2. Road supply and parking

Table 4.5 also looks at how cities provide for their transport modes in terms of road supply and central city parking. Here again the automobile cities of the US and Australia provide around three to four times as much road per capita as in the European cities and seven to nine times as much as in the three Asian cities. Central city parking does not have quite such a large variation with the US cities having some 80% more spaces per thousand workers than the twelve European cities and six times that provided in the three Asian cities. In Australia, Perth is the outstanding city for automobile provision, with by far the highest road supply per capita and central city parking provision (second only to Phoenix which does not have a true central city area).

3. Congestion and public transport speeds

Table 4.5 sets out the variations in provision for the automobile and the relative speeds of various transport forms. A very clear pattern distinguishing automobile-dominated cities from those with significant public transport use (particularly rail) is revealed by how easy it is to travel by automobile and how the transit option competes in time. The data in Table 4.5 show the automobile-based cities have average traffic speeds of 43 km/h (US) and 44 km/h (Australia) compared with the European cities of 33 km/h and the Asian cities of 24 km/h. On the other hand, the bus-only cities of the US and Australia provide little competition for automobiles with 21–23 km/h transit speeds. Only the rail option can compete with automobiles, as the average speed of urban trains is 42 km/h in the US, 45 km/h in Sydney, 43 km/h in Europe and 40 km/h in Tokyo. Tram speeds are much lower, but they act usually as distributors in central areas linking in to the major train stations (Vuchic 1981), and typically operate with very high passenger loadings especially compared with buses. It is also interesting that the average speed of buses in American, Australian and European cities as well as Toronto is 20 to 21 km/h, a remarkably constant figure considering the enormous diversity in urban conditions in these cities. In the very much denser and congested Asian cities, it drops to 15 km/h. It would thus appear that, in general, bus-based public transport systems seem to have an in-built limit on operating speed of no more than 25 km/h, and thus cannot be considered genuine competitors in speed to automobiles in any of these cities.

It could be concluded that any city seriously wishing to change the private automobile/public transport equilibrium in favour of the latter must move in the direction of rail-based systems. This is the kind of dramatic change occurring in Australia in Perth where its new electric rail service is proving to be a big boost to public transport. A similar dramatic change occurred in Brisbane with its new electrified rail. The question facing all Australian cities now is the extent to which they can consider a move to light rail as the means of boosting transit.

There is also clearly less gasoline use in cities with low average traffic speeds in Europe and Asia compared to the US and Australia with their high average traffic speeds. This contradicts those traffic planners who suggest freeing up congestion to increase average speeds will save gasoline (this debate is analysed in Newman and Kenworthy 1984; 1988a; 1988b). Although free-flowing traffic may improve individual vehicle efficiencies, the evidence suggests that it also causes overall fuel consumption increases, presumably due to greater private vehicle use. The evidence favours instead the policy of traffic calming which attempts to slow traffic down both as a means of managing local traffic impacts and facilitating modal shifts in cities. It deliberately sets out to provide for bicyclists and pedestrians on roads and to favour transit for longer distances. Australian cities are beginning to show interest in

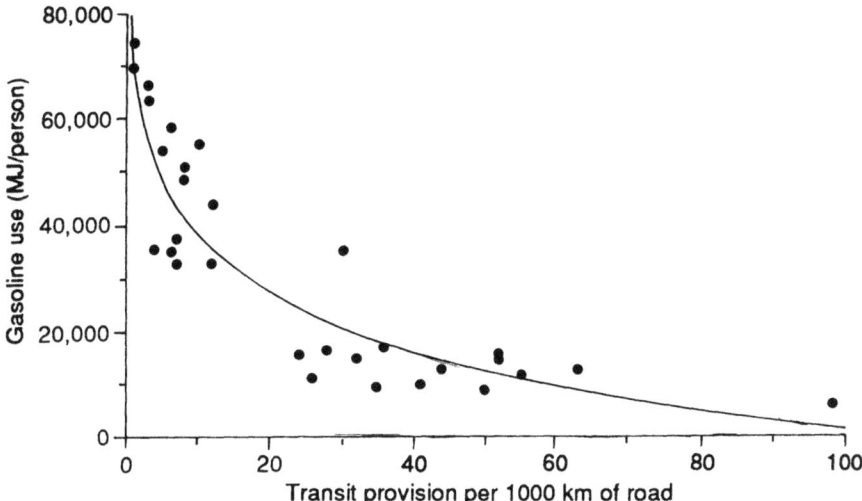

Figure 4.1 Gasoline versus relative transit provision in global cities, 1980.

traffic calming as a policy, although only Fremantle has done an area-wide traffic calming study.

The above data on relative infrastructure provision are summarised in Figure 4.1 which shows how gasoline use varies with the relative provision for transit over traffic measured by the length of transit service compared to roads. There appears to be a point at which the provision for transit becomes critically low, after which the city becomes heavily automobile-dependent and energy-intensive. Australian cities appear to be around the corner in the automobile-dependent zone but could relatively easily flip back with a change in infrastructure priorities.

The intensity of urban form

1. Total density

The main parameter describing the form of a city is its total density. The intensity of development in a city has a highly significant effect on travel distances and modal split (see Pushkarev and Zupan 1977). Table 4.6 sets this out in various parts of the city. The overall shape of the US and Australian automobile city is of low density in population and jobs with European cities generally being about four times more dense. Newer cities like Houston, Phoenix, Perth and Brisbane have densities around half that of the older cities like Chicago, New York and Sydney. Toronto tends to be more like a European city in its overall urban form, although

Table 4.6 *Intensity of land use in global cities, 1980 (per hectare)*

Average by cities	Whole city density: Population	Jobs	Central city density: Population	Jobs	Inner area density: Population	Jobs	Outer area density: Population	Jobs
Average, 10 US cities	14	7	54	500	45	30	11	5
Average, 5 Australian cities	14	6	13	360	24	27	13	4
Toronto	40	20	25	757	57	38	34	14
Average, 12 European cities	54	31	92	361	91	79	43	17
Average, 3 Asian cities	160	71	149	692	464	296	115	43

Source: Newman and Kenworthy (1989).

the data here refer just to the Toronto metropolitan area and not Greater Toronto. The Asian cities are again even more extreme with densities some ten times those of the US and Australian cities. Hong Kong is by far the highest density in the sample and probably the world.

2. Central city density

One of the significant differences between the US/Australian automobile-based cities and the more transport balanced European and Asian cities is that the former have central cities which have become areas of very high job concentration with generally few residents, and the latter have a much better balance between central city jobs and residences. The central city high rise office block is a common sight in most cities, but is a dominant characteristic of the automobile city and it gives US and some Australian cities (Melbourne has 647 per hectare) much higher average central city job concentrations than in Europe. In fact, the average job density profile of the US city is extremely sharp, going from 500 per hectare (ha) in the central city to 30/ha in the inner city and 5/ha in the outer areas, compared with European cities which have 361/ha, 79 and 17 respectively.

On the other hand the residential densities of US and Australian central cities are generally less than 20/ha (except for New York, Boston and San Francisco, which raise the US average from 15 to 54 per hectare), whilst in Europe they average around 90/ha and in Asia about 150. These patterns mean that public transport has some strong destinations in US and Australian central cities but highly scattered origins. With more sub-centres of job concentrations and more residential development in centres and sub-centres, there is the potential to reduce all travel needs and to ensure transit also has viable concentrations of activity. Canberra is a good Australian example of this policy, though it does not indicate much less automobile activity as the density of activity in its sub-centres is too low (Newman and Kenworthy 1991).

3. Inner city and outer area density

As well as being different in overall and central city density, there are clear differences in urban form between US/Australian cities and European/Asian cities in terms of their inner cities and their outer areas.

The US/Australian inner city is generally two to three times less dense than in European cities, and ten or more times less than that in Asian cities. However, the old inner cities of San Francisco, Washington, Boston, Chicago, Sydney and particularly New York are similar to many European cities, as is Toronto, while the inner city densities of new US

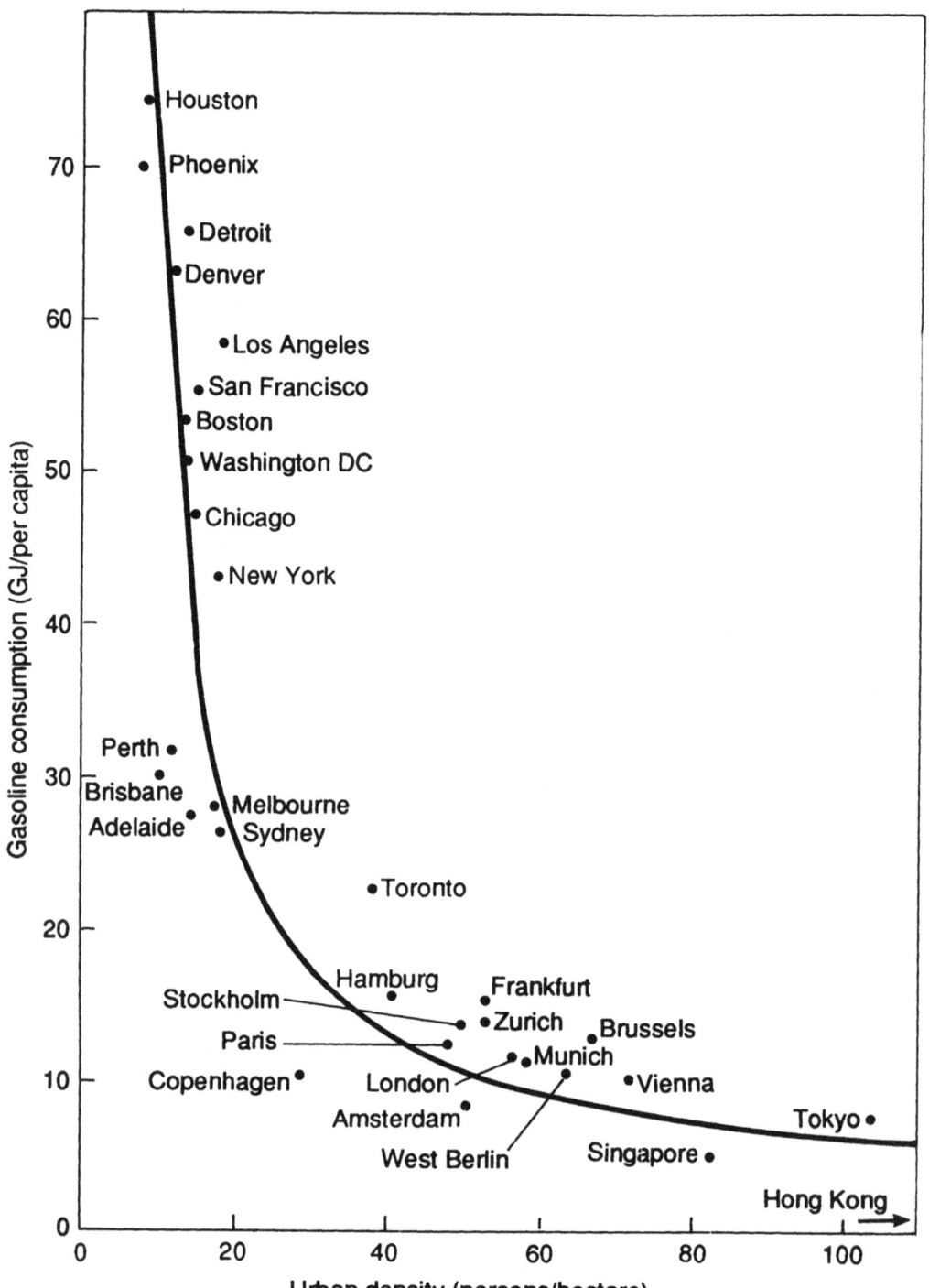

Figure 4.2 Gasoline use per capita and density in global cities, 1980.

and Australian cities are generally little more than their overall density. This confirms the generally accepted picture of older cities as having steeper population density gradients (e.g. Clark 1982). The automobile is obviously responsible for such dispersion of activity in newer cities and urban areas.

The outer area densities of US and Australian cities are amazingly uniform in all cases, with very low land-use intensity. The density of these outer suburbs is around the density of rural Java. This raises the question as to whether such areas should even be classified as urban in any traditional sense. Certainly the ex-urban areas of modern automobile-based cities are little different from rural areas. European cities are marked by much more intensively utilised outer areas, some four times more on average than in US and Australian cities. Thus even new urban areas planned in the era of the automobile are not nearly so dispersed in Europe as they are in US and Australian cities. Toronto is again more like a European city in its outer area; both Toronto and the European cities appear to develop their density partly through a number of intensively utilized sub-centres linked by rapid transit to the city centre.

Density patterns are obviously closely linked to transport and hence to energy use. This link to the intensity of urban development is summarised in Figures 4.2 and 4.3 which show the variation in gasoline use with density for the 31 cities, and public transport usage and density within Melbourne by local government area to indicate how the same

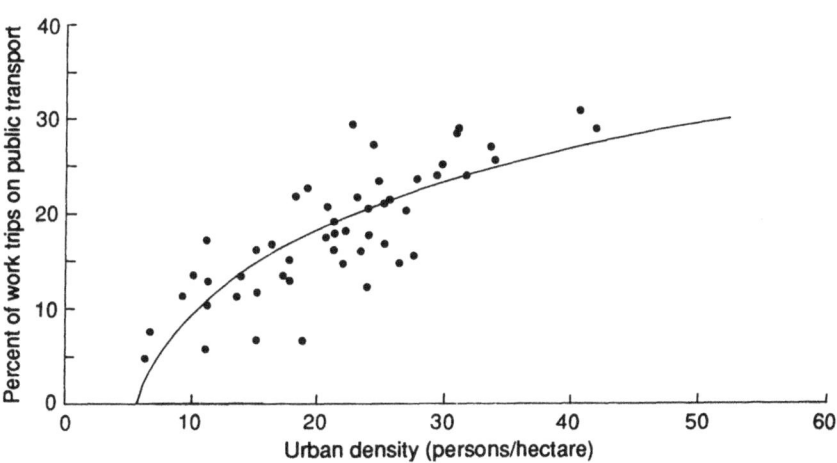

Figure 4.3 Public transport use for journey to work and density by local government area in Melbourne, 1986.

kind of variations in transport exist across a city. In both cases there appears to be a critical point (around 20 to 30 people/ha) below which automobile-dependent land use patterns are an inherent characteristic of the city. Thus there is support for increasing urban densities as a transport energy conservation policy. In particular, there is support for the kind of land use we have called *urban villages*, which is high density, mixed in jobs, houses and services, and linked to the rest of the city by good transit (Newman 1991; Newman *et al.* 1992). Such urban villages are a new design orientation in Australian cities and are being facilitated by the Commonwealth Government's Better Cities programme.

Policies for reducing transport energy use

The policies considered to be most relevant for saving fuel will be outlined based on the four main areas described above. They will be illustrated by reference to the Ecologically Sustainable Development (ESD) Transport Working Group (1991), and responses to this by the ACF–WWF ESD unit (Hare 1991).

Vehicle technology

The ESD Transport Working Group, which was one of the broadest-based groups in recent Australian history to consider transport and energy, considered the extent to which vehicle technology could be improved to save fuel. The conclusions from the Group are thus worthy of closer attention and are summarised in Appendix 1 at the end of this chapter.

The Australian motor vehicle industry is not going to be at the vanguard of any of these suggested moves unless strongly regulated to do so. They are committed to building a relatively large automobile as their niche market and do not see any significant vehicle technology improvements as coming from Australia. This attitude goes a long way to explaining the reason why the Sarich engine will not be mass-produced in Australia. This engine is now seen to be the most significant advance in engine technology for decades (Wright 1990). It is an opportunity lost for Australian industry to lead the way in a more sustainable kind of technology, and to use it as a means of making a more vital local economy in the process.

The conservation movement through the ACF–WWF considered that the ESD analysis did not go far enough, and saw significant potential through a more rapid move towards new technology in vehicles and to increased scrappage rates with vehicles. They argued as follows (Hare 1991):

- Voluntary fuel consumption standards are not supported.
- The Australian motor vehicle industry has already proven its reluctance to improve fuel efficiency, having announced a new vehicle fuel consumption target for the year 2005 weaker than the 1988 European average. Any targets voluntarily agreed to by the Australian automobile industry are likely to be so weak as to offer no real contribution to greenhouse gas emission reductions.
- Experience in the US demonstrates the effectiveness of mandatory standards in improving fuel consumption. An examination of US fuel consumption trends indicates that improvements over the last 15 years have been almost solely in response to State and Federal regulations, and would not have occurred otherwise (Goldstein *et al.* 1990).
- It is feasible for Australia to catch up by 1995 to the 1988 European average of 7.8 litres/100 km (a 14% improvement). This will require improvements in engine management, incorporation of world best technology such as lean-burn engines and a slight down-sizing of the fleet.
- Insufficient data are currently available on the fuel consumption of light commercial and recreational vehicles sold in Australia to allow specification of a standard for 1995. A survey is required to rectify this situation.

The ACF–WWF recommendations represent an alternative, stronger view and are set out in Appendix 2 at the end of this chapter.

Australia's vehicle fleet is one of the oldest on average in the OECD and is thus an issue in terms of fuel consumption. Whilst in one sense this may be good (product durability), older vehicles tend to be less efficient due to both technology and the ageing of vehicle components. This loss of efficiency was considered by everyone (ESD and conservation movement) to outweigh the benefits of durability at present. Very little action has occurred following this debate.

Prices

The thrust in Australia and indeed most parts of the world to adjust prices to fully reflect costs, including environmental factors, was another major consideration in the ESD Transport Working Group as well as other parts of the ESD process. The situation of Australia compared to other countries is shown in Figure 4.4, and the ESD working group recommendations for reform are summarised in Appendix 3.

The Working Group could not finally conclude anything much on fuel price increases such as a carbon tax. This was due to a strong case being presented for the equity aspects of such change. The poor are

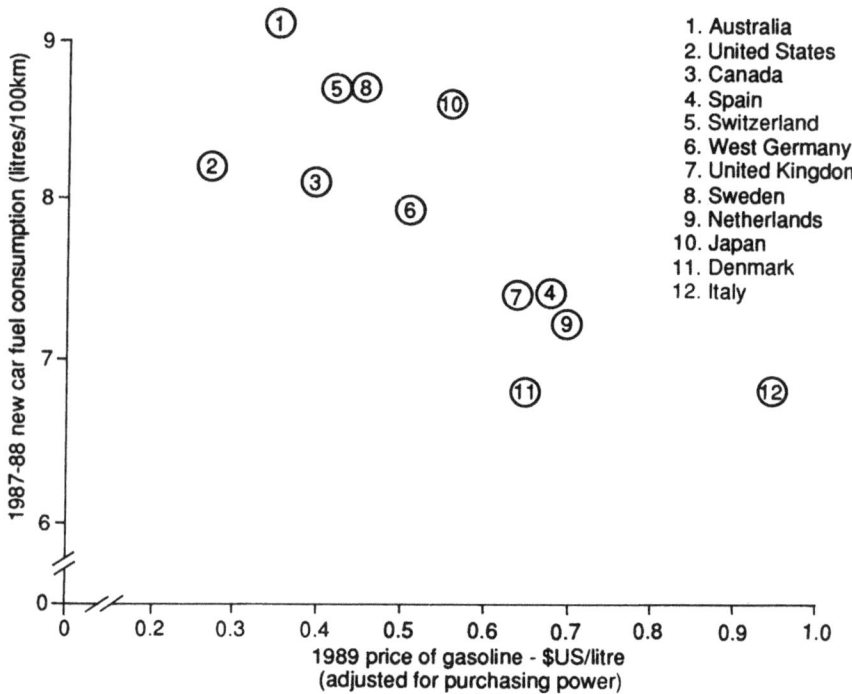

Figure 4.4 Gasoline price and new vehicle efficiency by nation.

increasingly located on the outer fringes of Australian cities, have little access to public transit, face large distances for walking or cycling, and so would be the ones bearing the brunt of any fuel price increase. The ACF–WWF argued instead for a small carbon levy which would not be designed to change travel demand in itself but to provide an incentive to reduce automobile use by providing capital for rebuilding Australian cities with new rail lines, traffic calming and transit-oriented urban villages. If public housing and social housing were a significant part of the urban village development then equity would be improved as well as reducing transport energy use. (See Chapter 8 for further discussion of carbon taxes, including distributional aspects.)

In the long term, increased fuel prices are inevitable as oil supplies dwindle and alternative fuels become necessary. In the meantime higher prices should be used to help prepare us for that time. Higher prices will influence all aspects of transport energy use including the structure of our cities and the efficiency of vehicle technology. (Figure 4.4 shows how fuel price relates to vehicle fuel efficiency.)

The other potential change if fuel prices increase is that there will be a move to more self-sufficient rural and country town land uses. Cheap fuel is one of the major causes for the decline of rural areas and growth of large cities – cheap private transport does not favour local processing, local shopping, local schools and other services. The potential for development of small towns and growth of innovations like permaculture villages is more assured in a future where higher fuel prices favour more local self-sufficiency (Newman 1991).

Since the ESD debate there has been no action on fuel price increases. Indeed, the only reaction was a proposal by the Federal opposition parties to reduce fuel taxes significantly, making Australian gasoline and diesel far cheaper. The implications of such a move, and in particular more automobile use, are generally only considered in narrow economic terms, with a highly ideological flavour. Indeed, the Opposition Leader, Dr Hewson, suggested it would be cheaper to scrap the Sydney and Melbourne rail systems and give commuters an automobile. A critique of this, including calculations that show Sydney would then need 11 six-lane freeways and 500 football fields of parking, is provided in Vintila *et al.* (1992).

Relative provision for transport modes

The ESD Working Group recognised the importance of transit in a more sustainable future and the relevance of traffic calming, bicycle planning and demand management in coping with the problems of automobile dependence. They recommended policies on alternative modes, demand management and infrastructure provision, as shown in Appendix 4.

Little criticism of these policies can be made. The ACF–WWF ESD response was to support them fully. The only question is the extent of commitment to them by governments in terms of infrastructure capital and other priorities. The central issue of how light rail can be introduced effectively into Australian cities was not addressed in the ESD process but is happening, partly due to the Commonwealth's Better Cities program and continued pressure from local communities (see Newman *et al.* 1992).

An important study by Glazebrook (1992) sets out a model for redeveloping Sydney around public transport, illustrating the energy, cost and land use advantages. Energy use in public transport is calculated to average 1.52 MJ/passenger kilometre over a whole day, while private transport is calculated to average 3.97 MJ. During peak traffic times, the difference is even more pronounced. A revealing comparison of network costs is given for a six-lane freeway against both double-track light and heavy rail. Both in terms of capital cost and cost per passenger km/hour, the rail systems show considerable relative advantages ($35,000/passenger km/hour for the freeway versus $800–$2100 for rail). The key to such savings is in the capacities of the three options, with maximum capacity for the freeway being 14,400 passengers per hour in both directions, against 24,000 for light rail and 35,000 for heavy rail (Glazebrook 1992). Finally, Glazebrook notes the amount of land used by the three systems, a calculation which also favours rail, particularly when stations and parking are included.

A choice must be made between the priority for roads and alternative means of transport. If the choice to focus on alternatives is made, savings in road construction can more than pay for the additional public transport and traffic-calming costs, and the infrastructure savings identified can be realised even after transport infrastructure is paid for. This is the theme of a scenario study on Australian cities (McGlynn *et al.* 1991).

Although public transport patronage will rise with reurbanisation, the total public sector operating subsidy to public transport will rise more slowly or not at all since a more compact form will allow more efficient use of capital, particularly if it reduces the directional imbalance in peak hour traffic. The greater priority given to public transport may also induce greater and more rapid productivity enhancing microeconomic reform in operations, with potential operating cost savings of the order of 36% (Travers Morgan 1991).

The intensity of debate on these issues in Australian cities continues to grow. Perth's new rail system has been extremely popular and has given a boost to the areas served with transit-oriented development being planned everywhere the lines run. This system was revived, rebuilt and extended due to a large scale community reaction that demanded a better future for the city than the Los Angeles model promised (see Newman 1991). Such community sentiments exist in most Australian cities.

Freight

The movement of freight uses significant quantities of energy in Australia, not surprisingly in a country of large distances and low overall population densities. There is little potential for improving the fuel efficiency of road freight vehicles, as this has plateaued following intense engineering changes in the 1970s. Thus the main potential for improvements in freight transport energy efficiency is via modal shift, that is, to shift towards rail freight, especially on inter-urban routes. To make this change would not occur simply, but would require considerable upgrading of the rail system. Much of the Australian inter-urban rail system has atrophied through lack of investment and use in recent decades, while the reverse has been true for road systems. Thus, a significant comparative advantage has been given to road freight over rail. In 1970–71, government rail and articulated truck freight tasks were 25 and 15 billion tonne kilometres (btk) respectively (Laird 1992a). By 1987-88, this had changed to 50 btk for rail and 60 for road. Rail freight uses significantly less energy than does road. While this varies with the technologies used, track gradients, distances and so on, rail may use less than half the energy to undertake the same freight task. Laird (1992a) notes the fuel savings which can be made if rail was improved to benchmark international standards and captured 70% of the modal share on major intercity corridors as being 200 million litres of diesel fuel per year.

The Commonwealth Government's 1992 'One Nation' package announced some investment in freight rail upgrading as part of the process of providing for a National Rail Corporation. However, it will be difficult to take freight off the roads whilst heavy haulage trucks continue to receive a subsidy equivalent to A\$5000 per vehicle (Laird 1992b). Any significant reduction in fuel prices, such as those proposed through taxation changes by the federal opposition parties in their 'Fightback!' statement in the lead-up to the 1993 Australian election, would make this modal shift virtually impossible to achieve, and could lead to the demise of the remaining interstate rail system.

The ESD Working Group (1991) report on transport recommended increased support for inter-urban rail freight, noting that the inclusion of the principles of sustainable development into policy considerations favoured rail. Energy savings are the advantage of rail freight discussed here, but there are other implications as well, including reduced pollution and road trauma, and spinoffs for inter-urban passenger rail (see: Australia. House of Representatives Standing Committee on Transport, Communications and Infrastructure 1989). An emphasis on achieving more energy efficiency in freight movements at this time in Australia is vital to maintaining a viable rail system.

These matters are examined further in a predictive model in the last section of this chapter.

The intensity of urban form

The ESD Working Group on Transport suggested a range of options on how changes in urban form could lead to reduced energy in transport. In particular it argued for the development of urban villages on transit routes as a central concept in reducing transport fuel. The final recommendations on urban form and transport are set out in Appendix 5.

Table 4.7 *Physical planning profile of Australian cities*[1]

Areas	Housing and urban form patterns: Urban density (people/ha)	Housing density[B]	Employment density (jobs/ha)	Land use mix	Transport patterns (to work) (%): Private car	Public transport	Walking, cycling
Core areas — Old city areas within walking distance of CBD — Heavily commercialised — Some high-rise housing	20–60 Av. 35	25–45 Av. 37	15–75 Av. 40	Very highly mixed, residential sparse in places (extremely self-sufficient)	52	32	16
Inner areas — De-industrialising — Gentrifying — Some redevelopment occurring so population decline reversing	25–40 Av. 32	10–25 Av. 19	5–20 Av. 11	Highly mixed in some areas; most have some mixture (high degree of self-sufficiency)	65	27	8
Middle suburbs — Ageing, occupancy declining — Beginning to decline in population	15–35 Av. 25	10–15 Av. 12	5–15 Av. 7	Pockets of mixed land use but mostly zoned (low self-sufficiency)	73	21	6
Outer suburbs — Developed — Young families dominant	10–20 Av. 14	5–15 Av. 10	2–7 Av. 4	Virtually no mix, highly zoned (little self-sufficiency)	79	16	5
Fringe areas — To be developed — Opposition from semi-rural inhabitants and environmentalists	5–15 Av. 8	3–10 Av. 5	1–5 Av. 2	Rural and some urban mixtures (some self-sufficiency)	84	9	7 (up to 18)

Notes: [A] Based on data from Melbourne 1981. [B] Housing density is gross residential dwelling density/ha (i.e. includes most roads).

Table 4.8 *Physical planning for Australian cities: housing and urban form targets*

Area of city	New housing developments (gross residential dwellings/ha)	Other land use
Core	High-rise housing mixed into other areas (100/ha). Facilitate housing above shops and offices in centres	Maintain major office and shop areas and community facilities in centres adjacent to transit
Inner	Redevelop deindustrialising areas at high-medium density (80/ha). Facilitate housing above shops and offices. Develop urban villages on large scale sites, especially near transit stations (80/ha)	Maintain major office and shop areas and community facilities in centres adjacent to transit. Local shops, offices, etc., to be mixed into urban villages
Middle	Allow as-of-right medium density (40/ha). Determine where housing stock could be redeveloped; facilitate by land assembly (at least 50/ha). Develop urban villages on transit lines (80/ha)	— as above for inner areas
Outer	Allow as-of-right medium density (30/ha). Remaining pockets of land to be at least Green Street densities (17/ha). Develop urban villages around present and new transit centres (70/ha)	— as above for inner areas
Fringe	Maintain rural zoning where possible. Development in concentrated urban villages (60/ha). Only develop other land if low impact and at least Green Street (17/ha), with transit vehicle access facilitated	Small offices and shops as part of urban villages

Table 4.9 *Physical planning for Australian cities: transport targets*

Area of city	Transit target	Traffic-calming target
Core and Inner	Fill missing links in public transport to ensure integrated rail–bus system and cross-city connections. Light rail in high use corridors, especially linking new urban villages. Improve frequency of services in short term, especially as densities increase. Higher priority to transit in road system. Improve transit environment, especially stations and stops. Develop community/ local government involvement and responsibility	Extend full pedestrianisation to more of CBD and selected inner streets. Develop public transport-only streets along key strip developments. Provide cycleways and wider footpaths where appropriate by reclaiming roadway. Curtail further loss of urban land to roads/parking areas
Middle and Outer	Extend rail to selected centres and urban villages. Timed-pulse integration of buses at rail interchanges and other significant centres. Provide bus only lanes on major highways and freeways wherever possible	Traffic-calm roads, e.g. near shopping centres, and improve design of major urban centres, to favour walking, cycling and public transport. Ensure cycleways/dual use paths focused on all major designations (e.g. schools, shops) via short, safe routes, both on-road and off-road. Reduce road widths and vehicle speeds in residential areas, enhancing sense of community and walking/cycling access
Fringe	New development built around transit, rather than reverse. Importance of feeder buses. Transit-based corridors with higher densities along routes part of fringe area planning. Community-based public transport with local responsibilities to become a feature of fringe areas (e.g. special buses)	New areas built with emphasis on traffic calming, through: lower proportion of land to roads; direct and shorter walking and cycling links to local facilities; greater use of grid-based system with physical techniques to reduce traffic speed and impact

The ACF–WWF response was fully supportive of these initiatives. Many of these reforms are quite difficult, however, and at grass roots level are often opposed by local environment groups. There is a great need to develop ways for local communities to participate in change that leads to a more sustainable urban form, and to identify local amenity improvements that can be part of that process.

Transit-oriented urban villages are now seen as a major design consideration in North American cities (see Calthorpe 1990; Rabinowitz *et al.* 1991a, 1992b). The Australian Better Cities programme is developing more than a dozen transit-oriented urban villages in all Australian capitals as a means of demonstrating this new way of improving our cities in a way that has inherently less transport energy (Newman 1992). We have also produced a set of guidelines for urban development and transport in each zone of Australia's cities that could achieve the simultaneous goals of reduced fuel use, greater economic efficiency and enhanced equity of access in an automobile-dependent city. A profile of Australian cities, based on Melbourne, is presented in Table 4.7. Matching physical planning targets are suggested in Tables 4.8 and 4.9.

Table 4.10 *Potential reductions in greenhouse gas emissions from Australian passenger transport, 1990–2005*

Measures[A]	Passenger task policy scenarios (MtCO$_2$)				1990–2005[B] (%)
	1990	1995	2000	2005	
Business as usual	46	49	53	56	122
Mild urban	46	48	51	54	117
Strong urban	46	46	47	47	102
Mild inter-urban	46	45	45	45	98
Strong inter-urban	46	44	44	43	93
Mild vehicle improvements:					
(a) 6.3 l/100 km	46	44	42	40	87
(b) 5.5% scrap rate	46	44	41	37	80
Strong vehicle improvements:					
(a) 4.7 l/100 km	46	44	40	35	76
(b) 3.5 l/100 km (including change to vehicle mix)	46	43	37	31	67

Notes: [A] Measures in the first column are introduced progressively: that is, each measure includes all preceding measures.
  [B] Minor discrepancies due to rounding.
  (For assumptions, see Appendix 6.)

Overall transport energy reduction potential

An overall response to transport energy conservation will have to address all the four areas of transport policy outlined in this chapter. This was the task set for the ESD Transport Working Group. As part of that exercise the ACF–WWF ESD Policy Unit developed a model to assess the potential of transport energy reduction through:

- urban system changes;
- inter-urban modal shifts; and
- vehicle technology improvements.

The model is expressed in millions of tonnes of CO_2 reduction as it was primarily oriented to the potential for reaching the 'Toronto' greenhouse gas emission reduction target. The model thus develops our work on cities and adds these other elements to it. The estimates for potential reductions through mild change or strong change policies are summarised in Table 4.10, along with the assumptions used.

What the model shows is that:

- continuing as we are will increase greenhouse gases from transport by 22% by 2005;
- urban measures could reduce this by between 5% and 20%;
- inter-urban changes could reduce business as usual by between a further 4% to 9%; and
- vehicle technology improvements could produce further reductions of between 6% and 26%.

Overall it is feasible to make reductions in CO_2 emissions of 19% less than business as usual, or virtually a return to 1990 totals, if mild changes are implemented. Alternatively, further reductions could be made: 45% less than business as usual, or 32% less than 1990 levels, if strong changes are implemented.

If Australian transport was to move towards the 'Toronto' 20% reduction scenario from 1988 levels, it would need to have a policy mix that was somewhat closer to the strong policy options outlined. The model does not directly include fuel price effects. As outlined earlier fuel price increases would appear to be essential if any of the above changes were to occur, particularly the more extreme 'strong' changes. It will also require highly directed policy changes implemented from specifically funded projects that can provide the necessary infrastructure, and a carbon tax seems to provide the mechanism for this.

Conclusion

To reduce transport energy use will require a combination of policies that address the need for improved vehicle technology, more appropriate pricing, a higher relative provision for transit and traffic calming in urban infrastructure, and a more intensive urban form based on urban villages. It appears possible that this could be achieved with economic advantages.

The kind of evidence outlined here suggests that at some point in the transition to non-automobile modes there is a range of factors that come into operation that make an automobile-oriented urban system very different from a transit-, walking- and cycling-oriented one. It is perhaps best understood in chaos theory by the way small incremental changes can bring about a sudden flip into an entirely new system. Only this can adequately explain the large variations in the two kinds of cities. This is a cause for great hope because, while it appears that little is happening to address the major concerns in the excessively oil-dependent cities of Australia and the United States, the beginnings of change are apparent.

The need for rapid action in the area of transport energy is highlighted by the considerations in this chapter emphasising city structure. We are now building cities that are likely to last at least fifty years; however, this is the period in which oil supplies will dwindle to negligible levels compared to consumption today. We will need to take energy conservation in transport very seriously.

APPENDICES TO CHAPTER FOUR

Appendix 1

Summary, ESD Transport Report conclusions on vehicle technology (*ESD Working Groups 1991*):

(a) that the Commonwealth Government, following consultation with the motor vehicle industry, ensure that a voluntary forward schedule is established setting the average fuel consumption of all new passenger cars, light commercial vehicles and 4-wheel drives sold in Australia, to be reduced progressively;

(b) that a regulatory mechanism be developed to be introduced if voluntary agreements fail to meet targets;

(c) that all new cars be labelled to indicate their fuel consumption rates as achieved under a properly audited test; and

(d) that all advertising in relation to the sale of new vehicles makes specific reference to the fuel consumption figures.

The close links between fuel use and emissions and their joint deterioration as vehicles age or are badly maintained were recognised in the following recommendations:

(a) that vehicle emission and fuel consumption testing facilities be maintained, developed or established;

(b) that more comprehensive new and tighter emission limits be introduced for diesel engine vehicles and for petrol engine vehicles;

(c) that the Australian Design Rules be modified to provide emissions specifications for vehicles designed to operate on other than petrol and automotive diesel fuel;

(d) that transportation fuel standards be developed and implemented that have greater regard to emissions that relate to the use of that fuel;

(e) that programs of random inspections of vehicles, to determine whether pollution control equipment is fitted and operative, be enhanced or introduced as applicable, and that penalties be imposed on owners of vehicles which are not properly fitted with operating pollution control equipment;

(f) that programs be enhanced and implemented, as appropriate, to detect and rectify vehicles emitting visible smoke;

(g) that a substantial Australia-wide study be undertaken into emissions from in-service vehicles;

(h) that if the project to assess the characteristics of in-service vehicle emissions indicates that there are substantial benefits in routine inspections of vehicles for compliance with air emissions requirements, an effective program be developed and implemented; and

(i) that procedures be further developed and implemented to control emissions from vehicles that are converted to run on different fuels, and that are modified from original design.

Appendix 2

Alternative ACF–WWF recommendations on vehicle technology (Hare 1991):

(a) a system of mandatory fuel consumption standards should be established for all classes of new motor vehicles;

(b) the standards would:
   – apply on a new vehicle fleet average basis,
   – be phased in from 1992 with full effect in 1995,
   – be structured with annual targets set over a ten-year period, with five-yearly reviews, and
   – specify in-use performance standards;

(c) the first ten-year standards for new passenger motor vehicle performance would be:
   1995   7.8 litres/100 km
   2000   6.2 litres/100 km
   2005   4.8 litres/100 km

(d) a system of in-use compliance testing be initiated, and linked to motor vehicle air emission compliance testing; and

(e) a survey be initiated immediately to obtain the data required to set in 1995 a fuel consumption standard for light commercial and recreational vehicles.

Appendix 3

Summary, ESD Transport Report conclusions on fuel prices (ESD Working Groups 1991):

(a) that concessions within the existing fringe benefit tax (FBT) system that encourage the provision of company cars be eliminated and that additional measures be taken for tax exempt bodies (such as government departments and agencies) to eliminate the over-provision and use of company cars, including eliminating sales tax exemption for governments in the purchase of new vehicles;

(b) that governments downsize their car fleets, improve fleet management, and encourage the use of public transport as part of salary packages;

(c) that sales tax on new motor vehicles of higher fuel consumption be increased and sales tax on new motor vehicles of lower fuel consumption be decreased, on a substantially revenue neutral basis;

(d) that the application of road pricing mechanisms be evaluated by Governments as an alternative to the provision of additional road facilities serving business districts and employment centres, and be considered as part of a program to enhance public transit;

(e) that the Commonwealth Government undertake a study into:
   –   how best to incorporate full economic, social and environmental costs into energy prices in Australian transport, and
   –   the merits and impacts of a carbon tax and a tradeable emissions scheme in reducing greenhouse gas emissions.

In regard to the second point the merits and impacts of a carbon tax should include attention to international competitiveness, equity issues, the advantages and disadvantages of earmarking funds to programs to support ESD and the potential for substantial price shifts in meeting greenhouse targets.

Appendix 4

Summary, ESD Transport Report conclusions on alternative modes, demand management and infrastructure provision
(ESD Working Groups 1991):

(a) that the Commonwealth Government and State, Territory and local governments identify potential for, and as appropriate implement, a suite of measures to reduce the demand for transport services within the community, and calm the flow of traffic in urban areas;

(b) that consideration be given by State and local governments to the provision of programs to require car and van pooling;

(c) where transit and road traffic are competing in a corridor or region, that, subject to the proper investment appraisal process (involving ESD considerations), transit and high occupancy vehicles (HOVs) should be given priority, such that transit time and travel time for HOVs is reduced;

(d) that traffic calming be given more prominence in local road planning and in the urban road funding process; that bicycling be facilitated as a growing part of the transport system and that, where appropriate, greater consideration be given to cycling in transport decision-making and planning;

(e) that a national cycle strategy which is integrated into national transport planning be developed and implemented;

(f) that priority be given by transit and local government authorities to the provision of dual-mode facilities for cyclists, through safe storage lockers, improved access to transit stations, and carriage of bikes on trains at off-peak times and where feasible in peak times;

(g) that an extensive program be funded to produce advice for municipal engineers and planners on travel demand management, traffic calming and bicycle facilities and to conduct training programs;

(h) that the full range of beneficiaries of urban public transport infrastructure be identified and that they make appropriate contributions to the costs of providing the infrastructure;

(i) that sufficient levels and sources of investment in urban public transport infrastructure be identified to assist in achieving appropriate modal shifts to achieve ESD objectives, and in particular should focus on areas of locational disadvantage;

(j) that governments and public transport authorities:
   – identify and implement measures to encourage greater patronage of urban public transport services, particularly where spare capacity exists, and
   – work towards improving the operational efficiency of urban public transport services;

(k) that to the extent indicated by investment appraisal procedures taking full ESD benefits into account, increased investment in inter-urban rail be undertaken, in conjunction with improved work and management practices to achieve best international standards of rail operational efficiency.

Appendix 5

Summary, ESD Transport Report conclusions on urban form (ESD Working Groups 1991):

(1) That a comprehensive program be initiated to increase the level of understanding of urban planners and managers and in the general community of the ways in which the social, economic, lifestyle and environmental benefits of higher urban densities and alternative urban forms in Australian urban communities can be achieved.

(2) That Australian cities should be developed more in the present urban area than on the fringe and with a range of housing types and densities; and:

(a) that subsidies on greenfield suburbanisation be phased out to achieve full cost pricing of urban development;

(b) that inappropriate regulations and processes preventing re-urbanisation be removed particularly those in the building industry;

(c) that the orientation of State Government land banking and land availability for new housing change from the urban fringe to redevelopment sites and to land acquisitions that can ease the problems of locational disadvantage in public housing and low income housing;

(d) that State planning agencies and local authorities conduct assessments of the capability for redevelopment in each local authority area through a detailed study of the infrastructure capacity;

(e) that local government desires for new development be integrated into effective metropolitan scale plans that provide clear zoning and priorities as to where and how much development should occur in inner, middle and outer zones;

(f) that particular attention be paid to redevelopment of land around key public transport nodes to include dense housing and some commercial activity so that travel is minimised and public transport facilities are made more inherently attractive and safe;

(g) that local authorities develop community consultative mechanisms to provide guidelines on urban design appropriate for redevelopment and to establish where the targeted housing could best be developed in order to achieve consolidation in their area;

(h) that innovative, affordable housing (as defined in the National Housing Strategy) be directed to locations that are close to sub-centres, to employment and to good transit access;

(i) that Commonwealth, State and Territory, and local taxation systems as they apply to housing be reviewed to determine whether there are ways:
- to provide for affordable housing more effectively (for example, higher density housing, co-operative housing),
- to assist in redevelopment rather than greenfield development, and
- to facilitate relocation for households wanting to shift closer to employment;

(j) that planning processes be developed that can provide more opportunities for precinct development (for example, urban villages) rather than for one housing lot at a time;

(k) that demonstrations of housing systems and urban villages be provided incorporating affordable and higher density housing at appropriate locations that can minimise or reduce travel;

(l) that principles of more efficient land utilisation developed by the Green Street Joint Venture be utilised where urban fringe development is necessary and that such development be provided as close as possible to transit services and to employment; and

(m) that dual occupancy and other small-scale additions to the building stock should be facilitated in established urban areas where large-scale redevelopment is not appropriate.

(3) That policies be finalised and implemented immediately to ensure that affordable housing for all housing groups (in particular for low-income households and groups in need) is in locations accessible to public transport, employment, and other essential community facilities, and that the need for travel is lessened.

(4) That new suburban employment be encouraged to focus as far as possible on public transport nodes in suburban centres, especially on rail nodes, and with access to medium-density affordable housing.

(5) That a study be undertaken into the ESD consequences of decentralisation.

(6) That existing urban transport and land-use planning arrangements be reviewed to achieve better integration of the planning system.

Appendix 6

Notes to Table 4.10: assumptions

Urban
1. Business as usual has an overall growth of 2.0% p.a. in passenger kilometres with automobiles and LCVs 2.0%, bus 1.3%, rail 0.8% and bike/walk 1.9%; bus, rail and bike/walk amount to 7.7% of total travel and average trip lengths increase by 13%.
2. Mild urban consolidation has 15% of new development in inner areas, 15% in middle suburbs and 70% in outer suburbs, with an overall p.a.

travel growth of 1.5% with automobiles and LCVs 1.5%, bus 1.9%, rail 2.1%, bike/walk 3.4%; bus, rail and bike/walk amount to 8.8% of total travel and average trip lengths increase by 6%.

3.   Strong urban consolidation has 100% of new urban development in inner suburbs or equivalent urban villages throughout the city with overall travel growth of 0.4% p.a.: automobiles and LCVs −0.3%, bus 4.2%, rail 5.3%, bike/walk 9.4%; bus, rail and bike/walk amount to 17.0% of total travel and average trip lengths decrease by 10%.

Inter-urban

1.   Business as usual is 2.2% p.a. growth in total passenger task involving annual growth rates of: automobile 2.0%, bus 1.3%, rail 0.8% and air 3.7%; public transport has 13.5% of total inter-urban travel.

2.   Mild modal shifts from automobile to bus and rail (50:50) producing annual growth rates of: automobile 0%, bus 5.1%, rail 11.8% and air 3.7%; public transport moves to 30.9% of total inter-urban travel.

3.   Strong modal shifts involve shifts from air to VFT-type rail and further automobile shifts so that growth rates are: automobile −1.0%, bus 6.2%, rail 16.1% and air 1.3%; public transport moves to 44% of total inter-urban travel.

Vehicle technology

1.   Business as usual is built into each of the above scenarios and is based on Australian Bureau of Agricultural and Resource Economics (1991) projections of 11% improvement in automobiles and buses, and 20% improvement in trains by 2005; the vehicle scrappage rate is assumed to be 3.5% p.a.

2.   Mild changes in vehicle technology are based on:
     (a) 14% improvement or 'catch-up' to reach 1988 European average new vehicle fleet of 7.8 l/100 km by 1995, through gains in engine management, world best technology (e.g. lean burn engines) and some slight downsizing, followed by OECD/IEA trend to 7.0 l/100 km by 2000 and 6.3 l/100 km by 2005; the scrappage rate is still 3.5% p.a.
     (b) the above technology improvements and a scrappage rate of 5.5% p.a.

3.   Strong changes in vehicle technology are based on:
     (a) 14% catch-up to 1995 of 7.8 l/100 km followed by a 60% improvement (6.2 l/100 km) to 2000 (based on OECD/IEA estimates of available technology, e.g. Sarich engine) and then to 4.7 l/100 km by 2005; alternatively, a less rapid rate of technology penetration and a bigger change in vehicle mix would be the same;
     (b) much the same technology changes but a bigger change to the vehicle mix thus going from 7.3 to 4.6 to 3.5 l/100 km; this would be achieved by government and company automobiles switching to small vehicles through skewed sales tax.

References

Archibald, R. and Gillingham, R. 1981. Decomposition of the price and income elasticities of the consumer demand for gasoline. *Southern Economic Journal.* 47(4): 1021–1031.

Australia. House of Representatives Standing Committee on Transport, Communications and Infrastructure. 1989. *Rail: five systems – one solution.* Canberra: Australian Government Publishing Service.

Australian Bureau of Agricultural and Resource Economics. 1991. *Projections of energy supply and demand, Australia, 1990–91 to 2004–05.* Canberra: Australian Government Publishing Service.

Australian Department of Primary Industries and Energy. 1991. *Road and transport energy.* Canberra: Department of Primary Industries and Energy.

Calthorpe, P. and Associates. 1990. *Transit oriented design guidelines.* Report to Sacramento County Planning and Community Development Department, Sacramento, California.

Campbell C.J. 1991. *The golden century of oil: 1950–2050.* Dordrecht: Kluwer Academic Publishers.

Chandler, W.U. 1985. *Energy productivity: key to environmental protection and economic progress.* Worldwatch Paper 63. Washington DC: Worldwatch Institute.

Clark, C. 1982. *Regional and urban location.* St Lucia, Queensland: University of Queensland Press.

Dahl, C.A. 1982. Does gasoline demand elasticities vary? *Land Economics.* 58(3): 373–382.

Ecologically Sustainable Development Transport Working Group. 1991. *Final report – transport.* Canberra: Australian Government Publishing Service.

Glazebrook, G. 1992. *Sydney at the crossroads: new land use and transport options for the future.* Draft discussion paper. Glazebrook and Associates.

Goldstein, D.B., Holszclaw, J.W. and Davis, W.B. 1990. *Efficient cars in efficient cities.* NRDC/Sierra Club Testimony for the State of California Energy Resource and Development Commission, Report on Transportation Issues.

Hare, W.L. (ed.) 1991. *Ecologically sustainable development: assessment of the ESD Working Group reports.* Melbourne: ACF–WWF ESD Policy Unit, Australian Conservation Foundation.

Kenworthy, J.R., Newman, P.W.G. and Lyons, T.J. 1983. *A driving cycle for Perth.* Final report to NERDDC. Canberra: Department of Primary Industries and Energy.

Kirwan, R. 1992. Urban form, energy and transport: a note on the Newman–Kenworthy thesis. *Urban Policy and Research.* 10(1): 6–23.

La Belle, S.J. and Moses, D.O. 1982. *Technology assessment of productive conservation in urban transportation.* (ANL/ES-130.) Energy and Environmental Systems Division, Argonne National Laboratory.

Laird, P. 1992a. *Australian road vehicle use and rail alternatives.* CRES working paper 1992/3. Canberra: Centre for Resource and Environmental Studies, Australian National University.

Laird, P. 1992b. Rational road user charges for heavy trucks. Paper to the Conference of the Australian Institute of Transport Research. University of Sydney, December 1992.

McGlynn, G., Newman, P.W.G. and Kenworthy, J.R. (1991). *Transport energy scenarios for Australian cities.* Report for the Sustainable Energy Future

Project for the Commission for the Future. In: *Towards better cities.* Melbourne: Commission for the Future.

Newman, P.W.G. 1991a. *The rebirth of the Perth suburban railways.* ISTP occasional paper 4. Murdoch, Western Australia: Institute for Science and Technology Policy, Murdoch University.

Newman, P.W.G. 1991b. Social organisation for ecological sustainability: towards a more sustainable settlement pattern. In: Cock, P. (ed.), *Social structures for sustainability.* pp. 27–42. Canberra: Centre for Resource and Environmental Studies, Australian National University.

Newman, P.W.G. 1992. *Sustainable cities: international and Australian progress: a perspective based on reducing automobile dependence.* Paper to the 2nd International EcoCity Conference, Adelaide, April 1992.

Newman, P.W.G. and Kenworthy, J.R. 1984. The use and abuse of driving cycle research: clarifying the relationship between traffic congestion, energy and emissions. *Transportation Quarterly.* 38(4): 615–635.

Newman, P.W.G. and Kenworthy, J.R. 1988a. The transport energy trade-off: fuel-efficient traffic versus fuel-efficient cities. *Transportation Research A.* 22(3): 163–174.

Newman, P.W.G. and Kenworthy, J.R. 1988b. Does free flowing traffic save energy and reduce emissions in cities? *Search.* 19(5/6): 267–272.

Newman, P.W.G. and Kenworthy, J.R. 1989. *Cities and automobile dependence: an international sourcebook.* Aldershot, Hampshire: Gower.

Newman, P.W.G. and Kenworthy, J.R. 1991. *Towards a more sustainable Canberra.* Murdoch, Western Australia: Institute for Science and Technology Policy, Murdoch University.

Newman, P.W.G., Kenworthy, J.R. and Robinson, L. 1992. *Winning back the cities.* Sydney: Pluto Press.

Newman, P.W.G., Kenworthy, J. and Vintila, P. 1992. *Housing, transport and urban form.* Murdoch, Western Australia: Institute for Science and Technology Policy, Murdoch University.

Pindyck, R.S. 1979. *The structure of world energy demand.* Cambridge, MA: MIT Press.

Pushkarev, B.W. and Zupan, J. M. 1977. *Public transportation and land use policy.* Bloomington, Il. and London, UK: Indiana University Press.

Rabinowitz, H., Beimborn, E., Mrotek, C., Yan Shuming and Crugliotta, P. 1991a. *The new suburb.* DOT-T-91-12. Washington DC: US Department of Transportation.

Rabinowitz, H., Beimborn, E., Mrotek, C., Yan Shuming and Crugliotta, P. 1991b. *Guidelines for transit-sensitive suburban land use design.* DOT-T-91-13. Washington DC: US Department of Transportation.

Travers Morgan Pty Ltd. 1991. *The effects of adopting international practice in Australia's rail systems.* Canberra: Industry Commission.

Vintila, P., Phillimore, J. and Newman, P.W.G. 1992. *Markets, morals and manifestos: Fightback! and the politics of economic rationalism in the 1990s.* Murdoch, Western Australia: Institute for Science and Technology Policy, Murdoch University.

Vuchic, V.R. 1981. *Urban public transportation systems and technology.* Englewood Cliffs, NJ: Prentice-Hall.

Wheaton, W.C. 1982. The long-run structure of transportation and gasoline demand. *Bell Journal of Economics.* 13(2): 439–454.

Wright, K. 1990. The shape of things to go. *Scientific American.* May: 58–67.

PART THREE

Renewable energy

CHAPTER 5

The nature of renewable energy

DAVID MILLS AND MARK DIESENDORF

Solar radiation already provides most of the energy which keeps the atmospheric, oceanic and biological systems of the Earth functioning. The *direct* solar energy that reaches the Earth completely dwarfs human use of fossil and nuclear fuels, being about ten thousand times larger. The sun also drives other atmospheric and oceanic activity which can be used. These *indirect* renewable energy resources are also large: worldwide wind potential is estimated at 2–3 times present world electricity consumption, hydro potential at about half present world consumption, and ocean thermal energy at about 200 times present world consumption. Our huge agricultural and biomass industries are solar-powered already, as are our own bodies (see Chapter 1).

This chapter briefly discusses the general characteristics of renewable energy relevant to the sustainability debate. The next two chapters review some of these technologies, considering the present status and future potential of these sources of energy. More attention is paid to those which the authors believe are more viable in the near term and which offer sizable energy resources. Longer-term prospects or smaller-scale technologies are noted briefly. Thus the emphasis here is on the immediate future; this is felt justified by both the apparent urgency of our need to reduce greenhouse gas emissions and urban pollution, and the very apparent profligacy in our present energy systems. This emphasis results in somewhat different focus than other reviews of the topic. General overviews of renewable energy options are provided by Grubb and Walker (1992), Rostvic (1992) and Johansson *et al.* (1993).

The range of renewable energy technologies and their current status are summarised in Table 5.1. Table 5.1 emphasises the large and diverse range of renewable options available. The rankings given to summarise status necessarily simplify a complex situation. It should be stressed that

Table 5.1 *Summary of renewable energy technologies*

Technology	Current technological status[A]	Current economic viability[B]	Potential long-term future contribution in Australia[C]
Solar heat:			
Various drying (salt, clothes, farm products)	1	1	1
Passive solar buildings	1	1	2
Domestic hot water	1	1	2
Residential heat	2	2a	2
Industrial heat	2	2a	2
Solar electric:			
Photovoltaics — flat panels	1	2b	1
— PV concentrators	2	2a	1
Solar thermal — parabolic troughs	2	2a	1
— paraboloidal dishes	2	2a	1
— central receivers	3	3	2
Other electrics:			
Hydro	1	1	2
Wind	1	1	2
Ocean thermal currents	3	3	3
Tidal	1	3	3
Geothermal	1	2a	2
Wave	2	3	2
Fuels:			
Biomass for burning (wood, bagasse)	1	1	3
Biomass for alcohol fuels (especially ethanol)	1	2a	2
Hydrogen	3	3	1
Energy storage:			
Batteries — standard	1	1	3
— advanced (e.g. vanadium redox)	2	2a	2
Rock beds	2	2a	2
Liquid salt, oil, etc.	2,3	3	2
Chemical reactions	3	3	2
Hot water storage	1	1	2

Notes:

A 1: Existing technology, commercially available (but not discounting further development).

2: Near-term technology for large markets (development to viable stage expected in next decade).

3: Longer-term prospects for large markets (more than a decade).

B 1: Economically viable now on a significant scale in Australia.

2a: Within 10 years of being viable on a significant scale.

2b: Economically viable now for small niche markets.

3: Not economically viable at significant or niche scales in Australia now, and not expected to be so within 10 years.

C 1: Large contribution (assuming it becomes economically viable).

2: Intermediate contribution.

3: Small contribution at best.

many renewable technologies are developing rapidly – their technological status may well change in the short term, and public policy settings may alter their economic viability. None the less, Table 5.1 provides a broad picture of the current situation for Australia at the time of writing. The three columns show the large differences in the status of the technologies. For example, conventional batteries are a mature, available technology that are economically viable now for suitable applications, but their contribution in the long term will always be small. In contrast, much more work is required to perfect the technology to produce hydrogen fuels, and for them to become economically attractive, but as a substitute for oil and gas, the potential contribution is very large. The various technologies in Table 5.1 are described in Chapters 6 and 7.

Why use renewable energy?

Human society already uses large quantities of renewable energy in various ways. As discussed in Chapter 2, traditional energy accounts only record hydro-electricity, wood and the small amount used by solar water heating. Usually, they ignore the very large amounts used in such activities as drying salt, clothes or agricultural products, and the passive solar heating of buildings. Inconsistently, these same statistics record the use of fossil fuels when this free and clean energy is replaced (e.g. electric clothes dryers or space heaters replacing passive solar).

At the broadest scale, renewable energy already contributes by far the greatest energy input to the maintenance of the Earth's dynamic systems. The rationale for using renewable energy as our prime commercial energy source is that we would be temporarily 'borrowing' a very small part of an existing clean energy flow, and therefore would be unlikely to perturb the global energy system. In operation, most renewable energy systems produce very little troublesome waste. On the other hand, the use of fossil fuels, while also small in relation to total global energy flows, produce waste products which can seriously disrupt other atmospheric processes and impact on the health of humans and other species.

Most renewable energy forms have little or no local effects on the environment during operation; possible exceptions are large dams, through inundation and water flow restriction, and the large-scale use of biomass. The latter may have adverse impacts from use of marginal land, extraction of nutrients from soil and, if the biomass is inefficiently burnt, creation of local pollutants. Whether global or local, the environmental and health impacts of operation of renewable energy systems are generally comparatively small, and the impacts of constructing the conversion systems are, at worst, roughly comparable with equivalent systems for fossil fuels. Because the embodied pollution during construction of either form of energy conversion plant is much lower than from the operation of fossil-fuelled plant, the total environmental and health impacts of renewable energy systems are very much less than those of operating fossil fuel systems.

There are concerns expressed about renewable systems based around aesthetics and land usage. Some have objected to the appearance of large installations, but these should be compared to the smokestacks, open-cut mines and waste dumps of conventional technology. Also, and contrary to popular myth, the land area used by most renewable energy systems to produce a given amount of useful energy is approximately comparable with that of a coal-fired power station together with its open-cut mines. Exceptions to this are again biomass conversion and large dams, which use larger amounts of land.

Data from the United States reported by Flavin and Lenssen (1991: 36) showed the following land occupation by different technologies (in square metres per GWh(e) for a thirty-year plant life):

coal   3642
solar thermal electric (STE)   3561
photovoltaics (PV)   3237
wind   1335
geothermal   404

The figure for coal includes mining, and that for wind is the land actually occupied by turbines and service roads. Land use estimates for nuclear power vary, but with fuel enrichment and waste disposal stages included are generally comparable to solar thermal. Some opportunities for dual land use also exist. Wind power plant can coexist with both pasture and crops. For solar photovoltaic systems there are ample rooftops; indeed, there is sufficient area on the roofs of Australian houses to generate more than the whole nation's electricity demand from solar energy systems without concentrators, with 8% conversion efficiency. The point of these figures is not to debate land usage, but to establish that the factor of land usage is not one of comparative disadvantage for renewable energy sources.

Nuclear energy has the advantage of negligible greenhouse gas emissions during normal plant operation, but is potentially a comparatively small, costly and inflexible resource with serious external disadvantages (see Chapter 2). Perhaps most seriously in environmental terms, it requires a large investment of energy and resulting emissions to construct such systems, and this could easily act to increase global warming rather than decrease it in a fast-growing implementation program.

It is difficult for society to agree on the cost of damage caused by polluting energy sources. One important role of research into renewable energy is to develop the technology to the point where the convenience and the cost of renewable energy is approximately equivalent to that of fossil fuel. It would then be sensible to abandon fossil fuel use because of its greater external environmental and social costs – a true 'no-regrets' option. Recent advances in renewable energy development are such that this task is close to completion in many important cases, and there are some situations where the cost of advanced renewable technology may soon be perceived as being less than fossil fuel systems without invoking such external costs. However, decision makers should not be afraid of attempting to calculate and include such externalities as urban pollution and global warming in the costing process. They are real costs which we all must pay.

There are other advantages of renewable energy. One is that many renewable technologies come in small modules, and this can be advantageous when there are large uncertainties in future energy demand. Small increments of investment are possible with renewables, and this allows flexibility and more rapid responses to emerging needs. Large increments in energy investment (e.g. coal mines, nuclear plants, etc.) demand better long-term predictions of demand than we are capable of, and are inherently inflexible because of high expense and long lead times.

There is also significant potential for employment generation. While changes in energy policy for environmental reasons are often opposed on the basis of threats to existing jobs, there are strong indications that investments in energy efficiency and renewable energy are more attractive in this regard than capital intensive fossil fuel systems. It has been suggested that investments in renewable energy and energy efficiency offer considerably more jobs per dollar invested, or per unit of energy produced or saved (Flavin and Lenssen 1991; Renner 1990).

Traditionally, a number of disadvantages have been associated with renewable energy, the primary ones being cost and the issue of storage. Cost is discussed in the next section, and in Chapters 6 and 7. The issue of energy storage is also addressed in the following pages, but deserves some overall comment. Because major renewables such as solar, wind, tidal and hydro power are not constant, it is often assumed that they can only provide energy, for example, when the wind blows or the sun is shining. These energy sources do fluctuate, but two important factors enable this to be overcome. The first is the presence of technologies that enable storage of heat or electricity, and these are discussed later. The second is the fact that, although fluctuating, renewable energies are not exactly random. The following points illustrate this:

- sunshine has a 24-hour cycle, modified by season and the presence of clouds;
- winds are predictable to a certain degree in all areas – some coasts have twice-daily seabreezes, and some inland areas have a midday peak;
- tidal power, if based on turbines which utilise incoming and outgoing tides, have four predictable peaks in output per day; and
- hydro-electricity depends on seasonal and year-to-year rainfall patterns, which we understand to a workable degree.

Thus, while subject to variation, renewable energies are based on natural systems that we have some understanding of, and can be modelled and predicted with reasonable precision. Particularly in mixed energy systems, and utilising energy storage techniques both available and in development, the issue of evenness of supply of renewable energy is not the insurmountable barrier that some people have supposed it to be.

Is renewable energy too costly?

Change in the energy sector in the past has been characterised by the division of decision-making into separated, uncoordinated market sectors, and is typified by poor involvement of the public in major planning decisions. Action which helps society as a whole will certainly increase costs in some areas, but this can be tolerated if the overall benefit is positive. The resistance to change by sectors with vested interests in existing technology and resources – such as the coal, oil, electricity and automobile industries – will be expressed as an opinion that new technologies are too expensive, even if the acceptance of such technology in the sector concerned reduces costs to society overall. Indeed, even technologies which would lead to immediate benefits in the sector concerned are obstructed, because powerful groups within the sector fear that they will lose their influence as their expertise becomes less attractive. Therefore, industry and the marketplace cannot provide the coordination necessary to minimise the cost of emissions reduction to society. Intervention, coordination and legislative targets are required from society at large through their representatives in government. The issue of cost is thus part of a larger issue of goals and priorities, and must be defined at an overall societal level. If one wishes to reduce greenhouse emissions and pollution, the Least Cost Planning method should be applied to the entire economy so that the overall cost of pollution reduction to the economy as a whole is minimised.

Least Cost Planning entails evaluation of the total internal and external costs of a planning decision, and the taking of effective measures to minimise these total costs. This implies knowledge of relative cost, pollution and economic factors across many sectors, and the ability to find the best course of action in a complex and highly technical system. It also implies the will to intervene in the overall energy market in a more determined manner than previously. Such intervention may be counter to some current economic ideology, but pure free market approaches have neglected consideration of the environment as an important economic problem, and the present market can often operate to create profit for some by redirecting external costs to others or to society as a whole. Such activities are not developmental, although they are often portrayed as such. True development increases the total wealth and well-being of society and the environment. Many current industrial activities may in fact reduce total wealth by using up or degrading natural resources and amenity. In the process, they often redistribute economic wealth from the many to the few, since the bulk of society usually pays for environmental degradation.

Many analyses are narrow, only considering simple economic measures to achieve Least Cost Planning. Use of the market to assist minimisation of the total costs of economic activities can be undertaken by the establishment of subsidies or 'user pays' systems so that external costs are reflected in the price paid by the user of the technology or product. In recent years, market-based approaches to environmental policies have become increasingly promoted but, as yet, not applied much in practice. These market interventions, however, must be enforced by legislation.

In fact, legislation is necessary to environmental improvement no matter what course one takes. The question is, what is the most effective approach? A carbon emissions or pollution tax is one approach, but not the only avenue. If one is not ideologically bound to market mechanisms, then simple legislation can be considered to enforce environmental goals. This is a common method used to achieve a certain outcome (such as reductions in traffic deaths or criminal activity), and also may be a more direct and less expensive method than using economic measures. One could legislate carbon emission targets required within a certain timeframe, penalise non-compliance, and allow the market to select the least expensive measures. An example is the fuel consumption (CAFE) legislation enforced by the Congress of the United States, a country which is the champion of free enterprise. This legislation is far from perfect and was heavily lobbied against and circumvented by the motor industry, but it achieved substantial environment gains, notably a doubling of passenger vehicle fuel efficiency between 1973–74 and 1979–80. Indeed, Greene (1990) demonstrated that 'the CAFE standards were a significant constraint for many manufacturers, and were perhaps twice as important an influence as price'. Greene concluded that 'the data are not inconsistent with the hypothesis that constrained manufacturers based their MPG (miles per gallon) product planning solely on the mandated economy standards'.

Economic evaluations often suggest that, in theory, regulation costs more than economic measures in achieving emissions reduction. However, there is considerable dispute as to whether this is the case in practice (e.g. Industry Commission 1991: D12). Suggestions that purely regulatory measures do not increase public welfare are usually evaluated on a narrow basis which ignores external environmental benefits, effectively disregarding the prime objective of such legislation.

Emission taxes can have implementation problems. Higher fuel prices raise production costs, affecting international competitiveness in those industries which use considerable fossil fuel. Governments can become 'hooked' on the income from pollution taxes, so tax thresholds which significantly reduce pollution may be resisted. Also, there is some question as to whether the economic cost of emissions can be estimated accurately, as this is the cost of a major atmospheric change or environmental calamity. There will always be disagreement between different parties as to such costs: a coal mining company will clearly present climate change as a less serious prospect than environmental activists will. However, many external costs appear to be so high that a policy to compensate for these costs rigorously may not be necessary, as well as not possible. The objective of any emission tax is, after all, the same as environmental standards legislation: to change the behaviour of the community, industry and the economy. Activities that produce emissions really only need to attract a sufficient tax such that new, cleaner technology is perceived to be cheaper in the market place. The question is: politically, can they be set high enough to drive producers and consumers into cleaner technologies? Faced with the above, pure legislative standards without taxes might be preferred. It may well be that a mixture of standards and taxes eventuates, such as is occurring in the United States petrol market. A mix of policy instruments is probably both the most likely and desirable course.

The Australian ESD process

Whatever the measures used, one must first develop agreed goals for total emissions, and then embark on choosing the most effective mix of policy instruments to achieve this while analysing the total impact of these policies. For example, legislation should not be simply introduced without an evaluation of the potential effect on the economy, particularly with respect to the timeframe for technological change. Most countries have done this very badly, with various interest groups using different bases for evaluation. Australia, however, has undertaken the Ecologically Sustainable Development (ESD) process, which used working groups made up of mixed interests such as business, unions, government officials and environmental groups. The working groups sought to establish basic data and evaluation methods, leading finally to negotiated consensus recommendations which were put to government for implementation (see ESD Working Groups 1991).

In the energy sector, the ESD Working Groups explored future energy scenarios to 2020 by using the MENSA (Multiple Energy Systems of Australia) computer model run by the Australian Bureau of Agricultural and Resource Economics (ABARE 1991). The complexity of a modern economy demands such an approach, but input data must be handled with care. It is most important that a modelling process with input data with wide community acceptance be established as a basis for future policy development. In the energy area, the Australian ESD process did this successfully. It is true that there were many deficiencies in the data supplied to the ESD economic modellers, but because what was established was a process, the modelling result can always be improved.

MENSA requires a database of the costs of various energy supply and efficiency improvement technologies. Using this data, the model then identifies an *optimum* outcome: the particular set of technologies which gives the minimum cost over the defined period, subject to a number of constraints. These constraints might comprise fuel availability, data about existing electricity transmission lines and gas pipelines and the costs of new ones, or, importantly, a firm greenhouse emissions target. The target of a 20% reduction in emissions by 2005, adopted by the Commonwealth government as an 'interim planning target', was the goal used in the ESD process. The model is useful because it identifies the relative value of actions to reduce emissions in different sectors, leading to a more informed set of policy priorities. It also shows what is possible in a given timeframe, provided governments introduce policies which allow the required technologies to compete in an imperfect market.

ABARE first constructed a 'business as usual' (BAU) scenario for the Australian energy sector, which assumes that technical and economic growth trends do not change. This scenario included existing inefficiencies of energy use, and so was not an economic optimum and could not be run on MENSA. Then ABARE ran two separate cases on MENSA:

1. an *unconstrained base case*, which included some cost-effective emissions reductions; and

2. a *greenhouse constrained case* which meets the year 2005 emission target of a 20% reduction from 1988 levels using energy efficiency, substitution of gas for other fuels, some domestic solar, wind generation, and some solar thermal electric (STE) generation without storage.

Table 5.2 shows relative CO_2 emissions for the Australian economy under the greenhouse constrained case. The model imposed a very large two-thirds reduction in 2005 electricity generation emissions from 1990 levels, and this is a major source of the high cost of the modelled electricity system between 2000 and 2005. It is the emissions reductions in the electricity generation mix which mainly reduced emissions in the commercial, manufacturing and residential sectors. About one-third of current electricity is used in the residential sector, and the program reduces emissions strongly in this sector because of low equipment lifetimes compared to other sectors.

The ABARE choice for the *greenhouse constrained case* (Figure 5.1) has energy levels 20.6% lower than unconstrained projections for 2005 at 409 PJ in the residential sector, but associated emissions are a very pronounced 66% lower. This is achieved not only using on-site energy efficiency, gas substitution for electricity and solar heat measures in the home, but also by the introduction of gas and renewable generation technology in the electricity sector to replace coal. Drops in other sectors mainly reflect emissions cuts from electricity supplied to those sectors. Thus, in preference to other measures, the program clearly determined that the coal-fired electricity sector was the main culprit in

Table 5.2 *Million tonnes carbon (MTC) emitted by sectors in ABARE greenhouse constrained case*

Source/contribution	1990	1995	2000	2005
Sector:				
Manufacturing	26.4	30.3	24.3	16.3
Commercial	8.0	9.5	6.4	2.9
Residential	14.1	13.6	9.3	4.8
Transport	20.9	23.2	23.7	24.0
Energy production	7.7	8.3	8.4	9.7
Total	**77.1**	**84.9**	**72.1**	**57.7**
Electricity sector contributions:				
Electricity (PJ)	495	553	545	524
Emissions (MTC)	38.7	42.2	27.9	12.1

Source: Adapted from Australian Bureau of Agricultural and Resource Economics 1991.

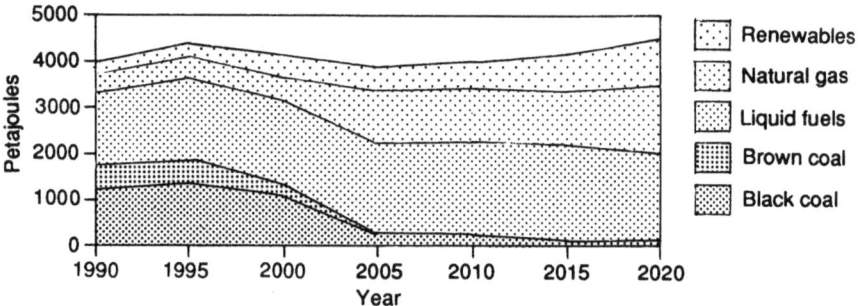

Figure 5.1 Total primary energy consumption by fuel, greenhouse constrained case.
Source: ABARE 1991.

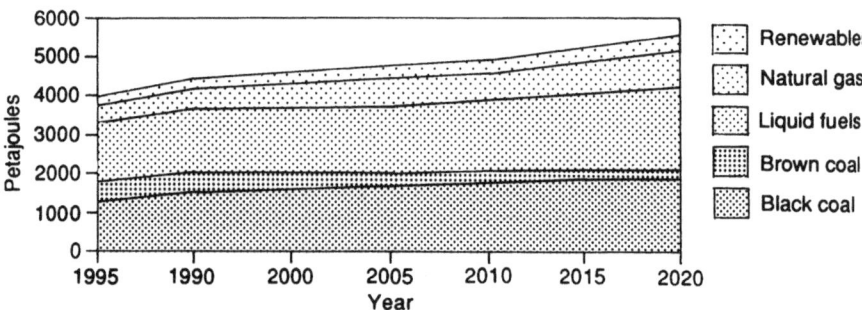

Figure 5.2 Total primary energy consumption by fuel, unconstrained base case.
Source: ABARE 1991.

environmental and economic terms, and under greenhouse constraints would act strongly to remove coal-fired generation. This is evident in the near elimination of brown coal in the early years of the next century. According to the model, while costs rise in the electricity sector, in total national terms the choice is an economic optimum given the data supplied.

The *greenhouse constrained case* estimated that of the total additional cost of meeting the year 2005 greenhouse target for CO_2 in the energy sector and maintaining it to 2020 was, without additional storage, about 4% of total energy sector costs, being some A\$ 40 billion or about 0.8% of projected total discounted value of GDP over the period.

Additional modelling showed that, with conservative estimates of current and future costs of electricity from the above sources, renewable energy (including existing hydro-electricity) without additional storage

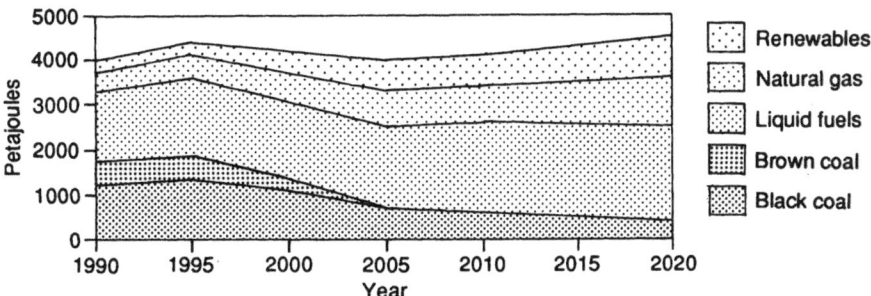

Figure 5.3 Total primary energy consumption by fuel, greenhouse constrained case including solar thermal electricity with storage.
Source: ABARE 1991.

could supply about 40% of electricity use by the year 2020. When hot rock storage is included for solar thermal electricity, this percentage increased to about 60%, and the cost of the national energy system dropped to within 2% of the *unconstrained base case* summarised in Figure 5.2. Thus, use of renewable energy actually reduced the cost of an energy system dedicated to meeting the 2005 emission targets compared to other technologies, and one such scenario is shown in Figure 5.3.

Could the cost of this different energy system be lowered further to meet 'no regrets' criteria? That is, would this different energy mix be potentially cheaper than what we would have done anyway, even ignoring environmental costs?

Reductions must be made in all sectors, but the residential market is a critical element in the initial stages of an overall carbon reduction strategy. Achieving adequate carbon reduction in this sector will make overall energy economy reductions less abrupt and more cost-effective. Homes in Australia are currently responsible for 30% of national carbon emissions, and make use of technology which can be changed over comparatively quickly.

Considering only the end use of energy in households, CO_2 reduction potential by 2005 in the domestic sector was estimated at 33% by Wilkenfeld (1990), compared to business as usual. By including more advanced solar thermal technology, Mills (1992) raised this estimate to 45%. There are several technologies available or potentially available that are between two and eighteen times more carbon efficient than direct electric heating. These are (Morrison and Tran 1992):

1. Moderate carbon efficient technology
    *   high efficiency reticulated natural gas or LPG gas space and water heating (2.1–2.3 times saving);

- solar water and space heating with electric booster backup (3–6.5 times);
- solar heat pump space and water heating (3–3.2 times);
- advanced domestic solar water and space heating with electric backup (7–9 times).

(The above includes a higher effectiveness group represented by solar/gas and advanced solar technologies which may be slightly more expensive to install.)

2. High carbon efficient technology
   - solar water heating backed up by natural gas or LPG (7.5–18 times);
   - advanced domestic solar water and space heating with natural gas backup (>15 times).

There is virtually no home which cannot use at least one of these technologies, and no drop in energy service standards is required. Solar can be easily used on two-thirds of homes, so it could be assumed that the 2005 technology mix can use two-thirds solar technologies either by legislation or by virtue of product lifetime financing programmes set up privately or by government. This reduces strongly the net investment required for CO_2 emission reduction in the electricity dedicated for the residential sector. Including the improvements in domestic solar technology in the model would drop the excess cost over the unconstrained base case to about 1% of the *unconstrained base case*, which, as we remember, is cheaper than business as usual. This is very close to a 'no regrets' program where purely economic criteria suffice to justify change, without considering environmental costs and benefits, and is probably within the limits of accuracy of any economic modelling program.

Yet even these costs are overestimates, since MENSA omitted some of the cheapest measures identified in the ESD process for reducing greenhouse gas emissions: reductions in energy demand through enhanced energy efficiency and conservation (see Chapters 3 and 4). During the ESD process the MENSA modelling was developed under high pressure, with the result that the incorporation of industrial and commercial sector energy efficiency and advanced transport technologies into the MENSA database was still in a rudimentary stage of development. Data on solar commercial and industrial heat were not included at all. MENSA allotted a larger than necessary task to changes in energy supply systems, which are by their very nature expensive to change. In particular, under the constraint of a greenhouse gas emissions target, MENSA tended to close down coal-fired power stations prematurely, thus greatly increasing costs. The MENSA modelling was also conservative in the sense that it did not include projected costs of energy technologies, which are currently expensive, but are likely to become cheaper in ten to fifteen years or more (see Chapters 6 and 7). These include solar photovoltaic cells, liquid and gaseous biofuels, and hydrogen storage.

Are there other alternatives, and in particular could nuclear power be substituted for renewable energy? A *nuclear baseload case* was also run with two costings. One was based upon revealed system costs in the United Kingdom for the Sizewell B system, and this was more expensive than the modelled solar baseload case. The other was based on CANDU plant costs estimated by the nuclear industry and was close in cost to the solar thermal electric/hot rock storage result (see Chapter 6). However, the model chose to derive more energy from renewable sources in this 'nuclear' model (mainly from wind and non-storage solar thermal electric generation) than from nuclear energy. The nuclear energy was used as baseload energy to support non-storage renewable electricity technologies. Clearly, the advent of technologies such as hot rock storage for solar thermal electricity could severely affect future prospects for baseload nuclear technology.

The high cost nuclear scenario was considerably more expensive than the solar baseload scenario. The low cost nuclear scenario proved slightly cheaper than the solar thermal baseload scenario, but costs were a subject of considerable debate. However, the cost debate may be irrelevant since the very high input cost of carbon-producing energy from other sources into nuclear systems during construction was not included in emission estimations. These would probably cause an increase in emissions rather than a reduction during the expansion phase of the program, so that the energy system would not meet the 2005 goals under any circumstances. The latest estimate for lifetime energy balance (energy out : energy invested) of a solar thermal plant is over 70:1 (Vant-Hull 1991), or between five and twenty times better than that of nuclear technology, depending upon the estimate used for nuclear.

Conclusion

The ESD results were very positive in that they clearly put the role of renewable energy into greater societal perspective, looking at overall benefits to the nation. Many of the potential benefits of renewables were not included in the ESD economic modelling, such as external environmental and health benefits, employment, electric grid related benefits, low energy investment, and low pollution investment and exports. Also, some renewable technologies were omitted entirely. Even given this, the 'solar energy-efficiency economy' came within 1–2% of the cost of the unconstrained base case. This suggests that such an energy system is desirable on economic grounds alone. A clean economy appears likely to be a cheaper economy.

It should be stressed that the renewable energy technologies required to achieve stabilisation of CO_2 concentrations in the atmosphere without significant change to lifestyle are available. They are in large part well understood and involve no great discoveries to be inferred, such as is the case with nuclear fusion. Renewables require no solution of intractable problems, such as the disposal of nuclear waste from nuclear fission and the neutron embrittlement of the core linings of nuclear fusion reactors. They would achieve a much better carbon emission reduction in the short term, because energy payback is much superior. Total energy reserves are far beyond what fission can deliver without use of the breeder cycle. Construction times are much below that of nuclear plant. This technical path simply involves refinement of what is much simpler engineering than nuclear alternatives, and then building up production to a large scale. This, and the ESD MENSA outcomes, clearly tell us that barriers to renewable energy development are not technical, but institutional and political.

These barriers are discussed further in the following two chapters (see also Chapter 9), but can be summarised here. The major barriers to the further development and/or adoption of renewable energy in Australia (and many other countries) are:

- a widely held but misleading perception that renewables can only offer minor or very specialised energy services: this is rooted in an ignorance of the range of renewable technologies that exist, and of the advances in this area in recent years; strangely, this misperception is highly apparent in the energy policy community;
- an inertia in the current energy system that operates in two ways: the reluctance of those with intellectual, political or economic 'capital' at stake to consider changes they perceive as threatening; and the more tangible problems of actively planning change and conversion of the energy system in an era when planning and intervention are unfashionable with many decision-makers;
- market and institutional distortions which actively operate against renewables: these appear either as significant incentives for fossil fuel-based systems (such as various subsidies to public electricity generation utilities), or as disincentives for renewable technologies (conversely, renewables receive no such protection and support, and are not provided with an adequate utility-like financing structure); and

- a tradition of lack of support in both the private and public sectors for research and development in renewable energy.

The issue of public funding for research and development deserves further comment. It is perhaps understandable that corporate leaders, operating under the immediate constraints and pressures of business, have overlooked renewable technologies; it is less justifiable that governments have done so. It was noted in Chapter 2 that renewable energy has been poorly supported on a global scale in this regard compared to fossil fuels, and in particular when compared to nuclear energy. In Australia, Kaneff (1990) has described the rapid decline of public funding for renewable energy research and development over the 1980s.

Despite these barriers, renewable energy clearly represents a stand-alone future, able to supply all needs of current and future human society. The next two chapters survey some of the available technologies in this area.

References

Australian Bureau of Agricultural and Resource Economics (ABARE). 1991. Costs of reducing carbon dioxide emissions from the Australian energy sector. In: *Economic modelling*, Chapter 5. Canberra: Ecologically Sustainable Development Working Groups.

Ecologically Sustainable Development (ESD) Working Groups. 1991. *Final report: energy production*, and *Final report: energy use*. Canberra: Australian Government Publishing Service.

Flavin, C. and Lenssen, N. 1991. Designing a sustainable energy system. In: Brown, L.R. (ed.), *State of the world 1991*. New York: W.W. Norton.

Greene, D.L. 1990. CAFE or price: an analysis of the effects of federal fuel economy regulations and gasoline prices on new car MPG, 1978–89. *Energy Journal.* 11(3): 37–57.

Grubb, M.J. and Walker, J. (eds). 1992. *Emerging energy technologies: impacts and policy implications*. Dartmouth, Aldershot: Royal Institute of International Affairs.

Industry Commission. 1991. *Costs and benefits of reducing greenhouse gas emissions*. Vol. 2. Canberra: Australian Government Publishing Service.

Johansson, T.B., Kelly, H., Reddy, A.K.N. and Williams, R.H. (eds). 1993. *Renewable energy: sources for fuels and electricity*. Washington DC: Island Press.

Kaneff, S. 1990. A decade of R&D decline. *Solar Progress.* 11(2): 5–6.

Mills, D.R. 1992. A no regrets solar energy strategy: an examination of pollution reduction assumptions in the ABARE/ESD MENSA model. *Solar Progress.* 13(3): 3.

Morrisson, G.L. and Tran, H.N. 1992. Energy rating of domestic water heaters. In: Australian and New Zealand Solar Energy Society. *Proceedings, Solar '92*, pp. 229–236. Caulfield West, Victoria: ANZSES.

Renner, M. 1990. *Jobs in a sustainable economy.* Worldwatch paper 104. Washington DC: Worldwatch Institute.

Rostvic, H.W. 1992. *The sunshine revolution.* Steingaten, Norway: Sunlab Publishers.

Vant-Hull, L.L. 1991. Solar thermal electricity: an environmentally benign and viable alternative. In: *Proceedings, World Clean Energy Conference,* Geneva, November 1991. pp. 350–356. Zurich: Cercle Mondial Du Consensus.

Wilkenfeld, G. 1990. *Greenhouse gas emissions from the Australian energy sector.* NERDDC end of grant report 1379. Canberra: Department of Primary Industries and Energy.

CHAPTER 6

Solar thermal energy

DAVID MILLS

This chapter reviews the potential for utilising solar thermal energy in sustainable energy systems. Two broad applications are examined: the production of electricity from solar thermal energy; and the use of solar energy to produce usable heat. More space is devoted to the former because, as explained in earlier chapters, electricity is a prime target for reform aimed at reducing energy usage and greenhouse gas emissions.

The solar resource

The first consideration in examining solar energy potential is the size and distribution of the solar radiation resource. Solar energy is widely and variably distributed over the surface of the earth. It can be collected either *directly*, using technology such as solar collectors and passive solar architecture, or *indirectly* from solar-driven biological, wind and wave sources. In this chapter we are concerned with direct collection.

In a good solar regime such as Australia, about 4.5–6.5 kWh of solar radiation will occur on each square metre of land surface each day ($kWh/m^2/day$). Figure 6.1 shows a map of average daily solar radiation striking a horizontal surface across Australia. (The map also shows the prime areas of wind energy potential, discussed in Chapter 7.) Generally, simple solar technology is suitable for areas having greater than 4.5 $kWh/m^2/day$, and high temperature solar thermal power plants for areas above about 5 $kWh/m^2/day$. It is clear from the map that direct solar energy is a huge and accessible resource for the Australian population. Moreover, towards the southern coastline where the solar resource declines, the wind energy potential increases, compensating for this.

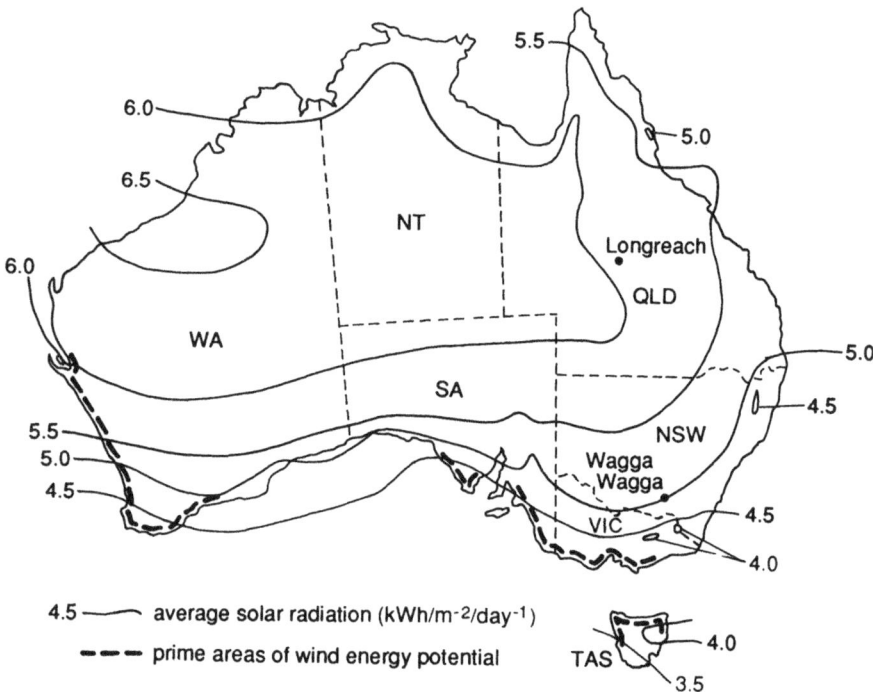

Figure 6.1 Global solar radiation (Wh/m²/day) and prime areas of wind energy potential in Australia. Longreach lies close to the 6000 Wh/m²/day (6 kWh) average insolation while Wagga Wagga is close to 5000 Wh/m²/day (5 kWh): these two towns are discussed as examples later in the chapter, and most of the Australian mainland lies inside these limits.
Sources: solar, adapted from Australian Bureau of Meterology 1975; wind, adapted from Blakers *et al.* 1991.

Solar radiation can be defined as comprising *direct beam* (arising from the solar disc), *circumsolar* (forward scattered), and *diffuse* (widely scattered) radiation. Some types of solar collector accept solar radiation from only part of the sky, and therefore fail to collect all of the available radiation. The central receivers and paraboloidal dishes described later in this chapter use high optical concentration to achieve high temperatures for some applications such as power generation. According to fundamental optical rules, such high concentration collectors have a narrow field of view and so collect mostly direct beam and less diffuse radiation. Low cloud areas with little diffuse radiation, such as deserts, are most suitable for such direct-beam-only collectors. Collectors which accept more diffuse and circumsolar radiation, such as low concentration flat plates and non-focusing concentrators, are more suitable for cloudier regions.

The total energy available to fixed collectors (which do not move to follow the sun's path) is approximately the same as the direct beam available to tracking collectors (which follow the sun's path), and may even exceed tracked direct beam in cloudier climates. This fact has important implications, as it broadens the technological options for different areas. Currently, the use of low concentration solar collectors is viable in temperate regions, while the traditionally favoured but more expensive high concentration tracking collectors are restricted to high solar radiation sites. However, some emerging improvements in technology that are described later in this chapter may make cheaper, low concentration collectors more attractive for high-temperature applications in both temperate and more sunny regions.

Solar thermal electricity

Solar thermal electric (STE) technology relies on the collection of solar heat at temperatures high enough to produce steam of a quality suitable for electric turbogenerators. This was difficult in the past, but is now routinely achieved. Conventional steam generation requires temperatures in the range 300–550°C. The operating temperature of current parabolic trough solar thermal electric plants exceeds 390°C, which is higher than in many nuclear plant boilers. Much higher temperatures can be achieved using paraboloidal dish technology, such as the 'big dish' developed at the Australian National University described later. Solar furnace temperatures of 3000°C have been achieved in France in a research facility in the Pyrenees (Trombe and Le Phat Vinh 1973).

Parabolic trough solar thermal electricity was experimented upon during the last century and used early in this century in Egypt for water pumping, but the increasing availability of cheap oil discouraged further development. Today, this approach to renewable electricity generation is less familiar to the general public than wind generation or photovoltaics (PV), but in spite of this STE is one of the most promising solar electricity technologies in terms of cost and availability. Current global generation from solar thermal plant is much higher than from PV plant, with over 350 MW installed in California alone, although this is still some seven times lower than globally installed wind-generated power plant at present.

The STE approach to electricity generation allows the use of conventional steam power generation equipment (that is, solar-heated steam turns a turbine which then turns a generator). However, such conversion equipment is complex, and this might seem to be a disadvantage when compared to the relative simplicity of photovoltaic (PV) systems. Indeed, PV would certainly be the better solar option for small or remote applications where small size and simplicity are very important.

However, for large-scale grid supply of electricity, it is STE that offers the largest and most immediate CO_2 emission reductions. The importance of the electricity sector in this sense was explained in earlier chapters. Before proceeding further, the following discussion will mention three types of electrical generation plant. *Baseload* plants produce the same output night and day, whereas *peaking* plants come on only for high-demand periods. *Load-following* plants combine these patterns, producing output which tracks the total demand profile of the power utility.

Solar thermal electric has several important advantages:

- apart from the solar collectors, the generating technology used is proven and familiar to electricity utilities, being almost identical to conventional plant;

- it appears that STE will be less expensive than large-scale PV for at least ten to fifteen years;

- STE can be used in cogeneration facilities, where the waste heat from the plant is used for local heating purposes;

- STE can combine well with fossil fuels, so that the same generating equipment can be used with either solar or fossil fuel derived heat. This means that a separate generating technology need not be used for low solar periods, with only supplementary fossil fuel being required;

- STE can be used with relatively inexpensive thermal storage to provide a twenty-four hour 'load-following' capability which allows replacement of conventional baseload fossil and nuclear plant. STE is fundamentally different from renewable electricity sources such as wind, wave and photovoltaic technology in that the renewable energy is first collected as heat, then converted into electricity. The others produce electricity directly. While the two-stage STE process might seem to be less advantageous, it in fact allows the storage of incoming energy in the form of heat rather than electricity. This is much cheaper than the electrical storage required by other sources; and

- energy payback time is very short. Vant-Hull (1991) provides interesting detailed information on the payback on energy investment of such plants, which will apply broadly to all advanced STE plants since they will use similar reflector field technology (the major part of the construction) and many will use the standard reheat steam turbine technology used in Solar Two (described below). He notes that the payback time in terms of thermal energy equivalent is about five months, or a 72:1 ratio of energy produced to energy invested for a plant with a thirty-year lifetime. This is better than current wind energy and future photovoltaic power (both about 20:1) and many times better than current nuclear plant. This means that with STE plant a rapid expansion of clean energy capacity is not going to compromise existing energy supplies and raise interim CO_2 emissions. This would be a serious, possibly prohibitive, problem with a rapid nuclear plant expansion.

Current STE technology

The largest and most effective current commercial STE projects have been installed in southern California by LUZ International Inc. (Jaffe *et al.* 1987). These plants utilise single-axis parabolic trough collectors that track the sun with a North–South axis of rotation. High concentration is used to reduce radiative losses to a few per cent, but the high concentration factor used also severely limits the collection of diffuse radiation, as diffuse radiation must be collected from a much larger angular segment of the sky than direct beam. The North–South orientation gives a high summer bias to annual electricity output, useful in California, but this reduces the total annual collection and gives poor winter performance.

The LUZ collectors are large. Large-diameter absorber tubes are necessary to allow adequate flow rates and pressures in the heat transfer fluid (oil), and this leads to the use of a large reflector. Each LS-3 collector has an aperture area of 545 square metres and uses 224 glass mirror segments. An 80 MW plant utilises 850 collectors. These are mounted parallel to the ground, making possible the use of big reflector units without heavy structures for inclination. In regions of low energy cost, the LUZ designs are currently not considered to be competitive, but LUZ estimated that the next major improvement in their technology would result in electricity generation costs of $US0.055 per kilowatt-hour in areas of high solar radiation. This is less than the cost of electricity generated from so-called 'clean' coal technology in the United States.

Another approach to the problem is the 'central receiver' design, often called the 'Power Tower'. These make use of large fields of reflectors which reflect sunlight to a common focal point receiver on the top of a tower. This has the advantage that a large network of pipes covering the array area is not required. Central receiver designs can reach very high temperatures, making them very suitable for some ancillary technologies such as gas turbine heat engines and high temperature molten salt storage. They have the disadvantages that reflector area is not used as efficiently as in parabolic trough designs, that diffuse radiation is not

collected, and that the reflectors must be individually computer controlled in two axes; each of several hundred or several thousand reflectors (called heliostats) has an individual tracking 'program'. Such
plants are also not very modular in nature: scaling up or down requires
redesign of the reflectors, receiver and tower, and large-sized plants –
greater than 100 MW(e) – would tend to be the norm.

The largest plant of this type is 'Solar One' at Barstow in California.
This plant ran succesfully for many years, but suffered from repeated
attempts by the pro-nuclear, pro-fossil-fuel Reagan Administration to
cancel the project. It is now being transformed into a more advanced
version which will become 'Solar Two' and which will use low-cost heliostats and a molten salt receiver. The molten salt also serves as a storage
medium providing about six hours of storage. Currently, the plan is to
construct, on the basis of experience gained with Solar Two, a large 100
MW(e) central receiver plant in the US by the year 2000. The electricity
cost from such a plant would be about $US0.11 per kWh for the first
plant and about $US0.096 per kWh for the seventh plant, which could
be cheaper than fossil alternatives in the US if environmental costs of
fossil fuel use were included (Vant-Hull 1991). Proponents of this type
of plant believe the cost could drop to about $US0.05–0.06 per kWh
with further improvement.

In spite of repeated statements by the US government that the central
receiver concept is superior, there seems to be little apparent cost
advantage in using the central receiver approach over the more flexible
advanced LUZ approach, although the provision of a storage system is a
distinct improvement from the perspective of an electricity generation
utility. The current Solar Two project at Barstow does explore storage
seriously for the first time, but it is questionable whether the molten salt
storage system used is competitive with alternatives such as hot rock
storage, which could be used with systems of the LUZ type. The issue of
energy storage will be discussed later in this chapter.

Despite the success of the LUZ technology, current solar thermal
daytime peaking plants are limited in potential. Ultimately, we will need
more than expensive peaking electricity to truly supplant polluting
fossil fuel energy supply technologies. To significantly reduce CO_2
emission levels, we will require lower-priced electricity available on a
twenty-four hour basis. Only in this way can we remove fossil fuel as the
foundation of our electricity supply. At suitable sites, wind power can
supply such baseload power, but is unlikely to contribute more than
20% of annual electricity generation from a grid without expensive
storage or backup plant.

What are the prospects for inexpensive STE baseload power? Is there
an option better than current technology? In the following sections we

examine these questions, and explore prospects for generating the majority of grid supply electricity from solar thermal energy.

Advanced solar thermal collector technology

While the LUZ systems represent a tremendous technological advance, they have descended from 1970s technology and there are now a number of ways in which cost and performance could be significantly improved. Two of the most promising low-cost advanced technology options are being developed in Australia.

Two-axis tracking (point-focus) collectors

One technology is the paraboloidal dishes ('big dishes') developed at the Australian National University (Hagen and Kaneff 1991). These could collect about 28% more energy per unit of collector area than LUZ technology because they exhibit very low losses and face the sun directly at all times. It seems that this 'high concentration, large reflector' approach offers a roughly similar generating cost of electricity – as low $A0.07 per kWh – as the next generation LUZ or two-axis tracking dish technologies. Point-focus technologies are most suitable for use in very sunny climates such as central and northern Australia, where direct beam is high, and diffuse and circumsolar radiation is low.

Small polar-axis tracking trough collectors

These are similar in overall appearance to the LUZ collectors, except that the the axis of rotation is tilted more upright, at an angle equal to the latitude angle from vertical. This inclined, polar tracking configuration actually gathers 97% as much annual solar radiation as a two-axis tracking dish of the same aperture (before thermal losses), yet utilises simpler engineering.

Previously, this low concentration technology could not be used for high temperature applications because thermal losses were too high. The advantages of polar-axis tracking can now potentially be exploited in much smaller collectors (about 2 m^2) with the assistance of new low-loss surfaces for use in solar evacuated tubes, developed at the University of Sydney (Dayton 1991; Zhang and Mills 1992). Although solar collectors utilising these new selective surfaces will not be available in demonstration quantities for at least three years, it is possible at this time to estimate the performance of collector systems using this new technology.

The use of low-loss selective surfaces at elevated temperatures would allow a concentration of only four to eight times, rather than the much higher twenty-five times that LUZ utilised. This means that the parabolic trough mirrors can be considerably smaller, and allows exacting engineering standards to be relaxed, thereby reducing costs. Small reflectors can also be more easily mass produced. In addition, low

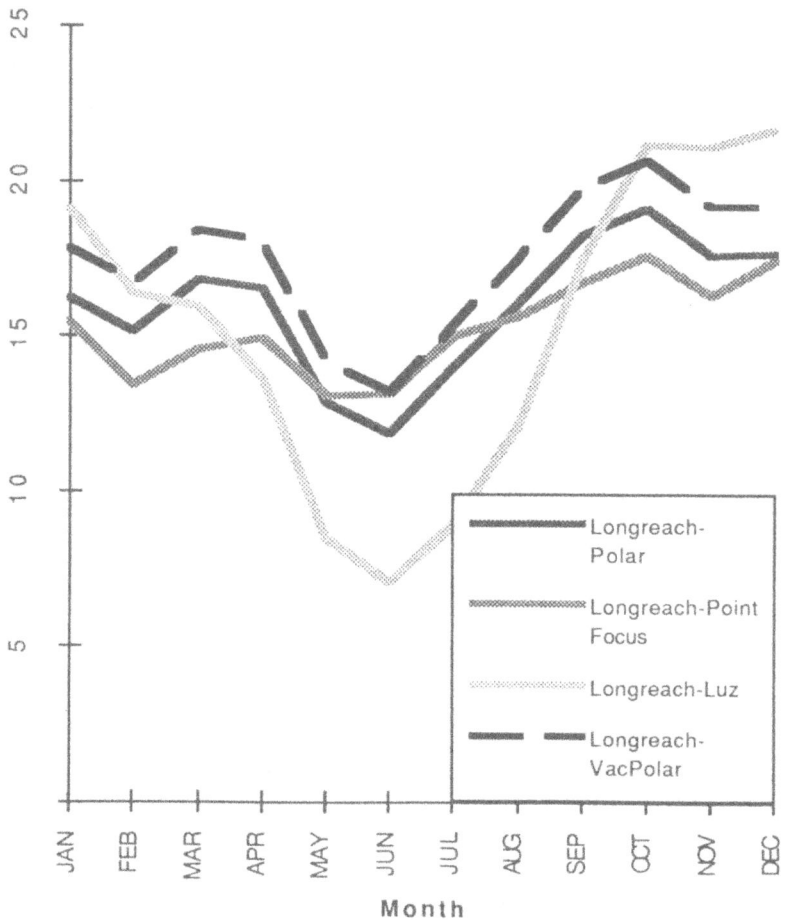

Figure 6.2 Calculation of daily thermal output of three collector system types in a high insolation region, Longreach, Qld. The polar collector performs best, and relatively consistently year round, but is only 10% better on an annual basis than the LUZ system, which is optimally oriented for summer at this latitude. Performance can be substantially improved by using vacuum insulated steam conduits (shown for the polar case).
Source: calculation by A. Monger and D.R. Mills, Department of Applied Physics, University of Sydney.

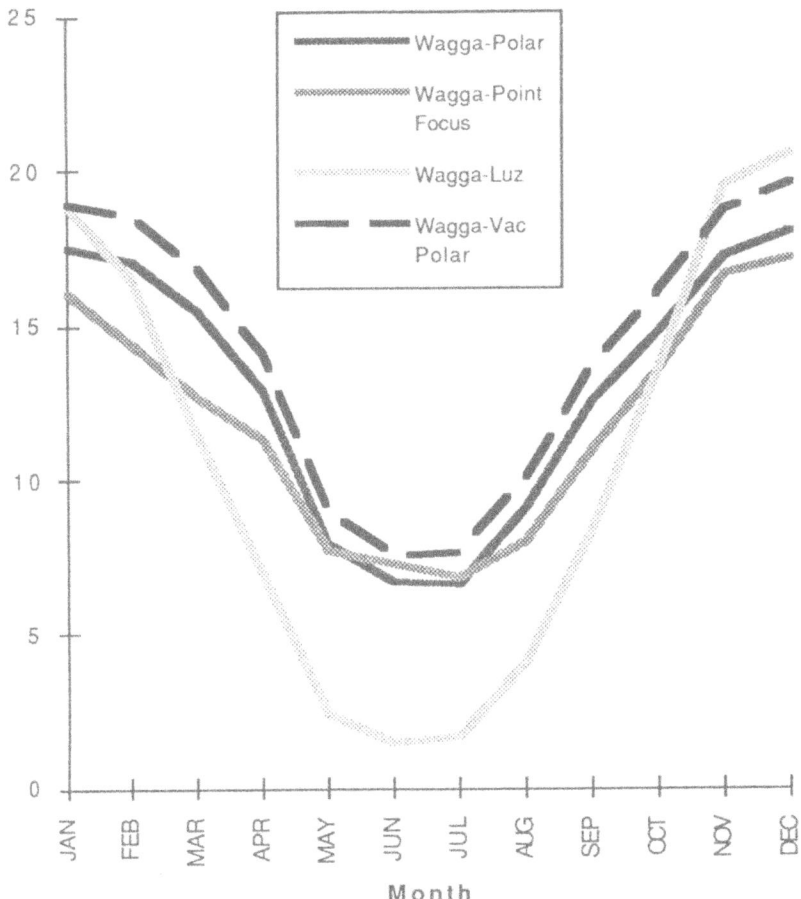

Figure 6.3 Calculation of daily thermal performance of three collector system types for Wagga Wagga, NSW. The LUZ system performs less well in this climate and latitude, with very little winter output. The polar trough collector performs best, being 24% better than LUZ because of improved diffuse acceptance, and again both polar and point-focus types can be substantially improved by using vacuum insulated steam conduits (polar being then 36% higher than current LUZ performance). Considerable gas backup would be used to maintain system performance in winter, but as shown later this will not add greatly to overall costs.
Source: calculation by A. Monger and D.R. Mills, Department of Applied Physics, University of Sydney.

concentration design allows use of much improved and simplified engineering to reduce costs. For example, the University of Sydney has pioneered an inexpensive passive boiling heat transfer system that requires no pumps and uses inexpensive common sizes of tubing. The system is only possible using low concentration optics and is not suitable for higher concentration systems such as LUZ, because flow resistance in small absorber pipes is too high for passive circulation. Also used are simplified low precision tracking, and a design which allows components to be changed by one person without shutting down the array.

Modelled performance results are shown in Figures 6.2 and 6.3 for different collector types at two representative sites in Australia for which detailed solar radiation data are available; Longreach in central Queensland and Wagga Wagga in southern New South Wales. Because of higher diffuse radiation collection, the small polar collector performs relatively better in poorer solar regimes, and thus would be more suitable for widespread implementation. Point-focus dishes and polar troughs are roughly comparable in high insolation regimes, with troughs having an edge at 350°C. At much higher collector temperatures, the point-focus collector would be the better choice, having substantially lower thermal losses.

It may seem from this that poorer climates like Wagga Wagga would be much less suitable for solar thermal technology. However, as will be shown later, the economics actually prove to be rather similar if the plant is backed up by a convenient fossil fuel such as natural gas. Figures 6.2 and 6.3 show relative performance in a good solar regime (Longreach), and a poorer one more typical of the major population centres in south-eastern Australia (Wagga Wagga). Winter solar collector performance is considerably reduced in the latter, but there is more diffuse radiation present so that the low concentration mini-polar collector performs relatively better than the others. The LUZ collector performs poorly in this situation, with low winter performance. At higher collection temperatures, the point-focus design would be more favourable, although thermal losses in field piping would be significant and the performance would drop from that shown in Figures 6.2 and 6.3.

Hybrid STE/fossil fuel plants

There has been a tendency in most STE research and development programs to provide a solar collector which has a comparable output temperature to that of the turbine used, with high superheat conventional turbines running at 550°C and the lower superheat LUZ turbine running at 390°C. The LUZ, the ANU Big (paraboloidal) Dish (Hagen and Kaneff 1991) and the Solar Two projects are all designed to be able to run a turbine with solar alone at certain times a day; that is, without any fossil fuel being used. This has tended to bias development in favour of high concentration, high temperature (and therefore high cost) solar designs.

However, it can be shown that a moderate amount of fossil thermal energy added to the solar output can superheat the saturated steam output of a 350°C solar collector to these levels, allowing the solar energy to be converted at higher efficiency through a high temperature turbine at temperatures up to 540°C. Thus a less expensive, low temperature collector technology can be used at the price of a small fossil fuel penalty. This is a 'hybrid solar/fossil fuel' approach. It utilises the natural high temperature advantages of fossil fuel to best advantage while reducing total fossil fuel usage very substantially. Fossil fuel is not only a backup in poor weather, but also an integral part of the energy supply at all times. Solar energy collection is increased because the collector field can run at a lower temperature (all thermal solar collectors run more efficiently at lower temperature) and solar energy collected is converted at a higher efficiency.

Although this might seem a compromise in terms of CO_2 reductions, it is superior to the traditional 'solar fuel saver' approach which saw solar merely as a minor input. This is illustrated by hypothetical computer modelling run for the Pacific Gas and Electric electricity grid in California carried out at the University of Sydney. It assumed complete grid supply by a 350°C collector system using a turbine at 550°C, and showed that about 82% of energy could be supplied by solar. This included backup in poor weather (Mills and Keepin 1993). Given that the backup fuel would likely be natural gas, which has 66% of the carbon emissions of coal per unit of energy output (on a typical mix of Australian coal used now – Wilkenfeld and Energetics 1991: 66), the result would be a system with emissions only about 12% of the CO_2 emissions of a purely coal-fired grid system.

Thermal storage

A common argument against solar energy is that it cannot produce power when the sun is not shining, and hence that there is no solar equivalent to conventional baseload power plants. There is some merit to this argument; peaking energy from, for example, a photovoltaic or solar thermal array, would require a large – usually fossil-fuelled – baseload system to underpin the system. Overall CO_2 emissions from such a system would remain high.

Currently, hydroelectricity is the only renewable electricity source able to supply energy cheaply and continuously on a twenty-four hour basis. Hydro is, however, limited in availability in many countries, particularly on the dry Australian mainland with its highly variable precipitation, and so is mostly used to supply power during peak demand periods. STE is potentially a much larger source of energy, and can also be adapted to deliver energy on a twenty-four hour basis. Thus, it is quite possible that STE could be the foundation of any future renewable electricity supply system in countries with a suitable climate. One way to accomplish this is with thermal energy storage, which has the potential to allow for precise matching of electricity supply to demand at reasonable cost. While much work has been done on energy storage at low temperatures, less effort has been devoted to high temperature storage. This is surprising, as thermal storage is likely to be much cheaper than electrical or mechanical means of storage (batteries, fuel cells, flywheels) which receive considerable research funding.

In the Solar One central receiver plant, an attempt was made at thermal storage using high temperature heat transfer oil circulating in rock. Such oil is expensive and the system developed a number of problems. In Solar Two, the storage system will use molten salt which is also used as a heat transfer medium from the central receiver. The storage medium here is more suited to the high temperature of the plant, but there are still potential problems – the system may freeze, and costs are still high.

There is a simpler approach. Rock beds have been used as reliable solar storage at low temperatures for many years (Swet 1990). Rock bed storage involves a large container of suitably sized rocks, into which heated air from a solar collector is introduced using a fan for circulation to all parts of the bed. Energy is passed to, and withdrawn from, the bed through heat exchangers. Early studies by the United States Solar Energy Research Institute (SERI) suggested that rock beds could be an economical method for large scale, high temperature storage with duration longer than a few hours (Dubberley et al. 1981a, 1981b). However, SERI was partially shut down during the Reagan administration and this work did not continue. At the Universities of Sydney and New South Wales, a currently confidential engineering study suggests that an extremely simple, high temperature rock bed heat storage system would be cost-effective (Mills and Morrison 1991).

Why have not rock beds been used up to now? Perhaps they are too simple to be interesting to researchers; the physics and engineering are mundane. A recent phase-change material storage conference agreed that heat storage systems like rock beds could be viable energy storage systems for power plants, and that they seemed to be much cheaper than the other options being discussed (Solar Energy Research Institute 1988). Yet nobody seemed interested in actually building such a system. Perhaps what is required is a strong expression of interest by electricity utilities so as to bias storage research activity away from the esoteric and toward the practical.

Grid connection aspects

The great potential of STE technologies is their suitability for supplying power to major electricity grids. So the question of connecting STE to existing grids is important. The preceding discussions of solar/fossil combinations and storage indicate some encouraging developments. Also very important is the matching of solar supply with electricity demand. Figure 6.4 shows the net system load (electricity demand) pattern for the Pacific Gas and Electric utility in California. If solar thermal, or photovoltaic, systems were utilised to supply as much of the load as possible without storage or natural gas backup, the collector system output pattern would look like Figure 6.5. Attempting to utilise any more solar plant would not reduce the residual peak loadings, which

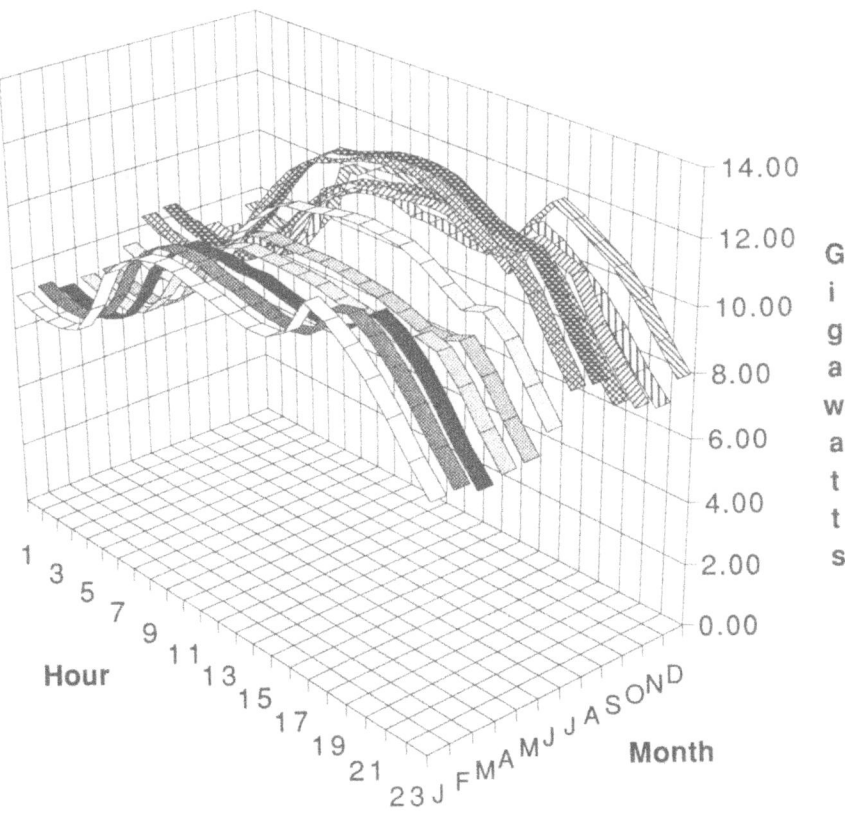

Figure 6.4 PG&E utility load pattern as a function of time of day and season. *Source:* Mills and Keepin 1993.

occur in summer evenings, and only 13% of the total load could be supplied without replacing cheap baseload energy. Peaking solar may be able to supply more than indicated by this example, if current daytime 'peak-lopping' demand management practices were to be reversed; a solar utility would possibly seek more demand in daytime hours. Nevertheless, it is clear that in order to achieve the full energy and CO_2 emissions reduction potential from solar thermal energy, at least some of the output from solar plants must be available around the clock.

For the utility load pattern in Figure 6.4, we can alternatively use a thermal storage system that stores about one third of daily collected energy. The storage volume chosen is based on a recent analysis of storage with natural gas backup at the University of Sydney, which

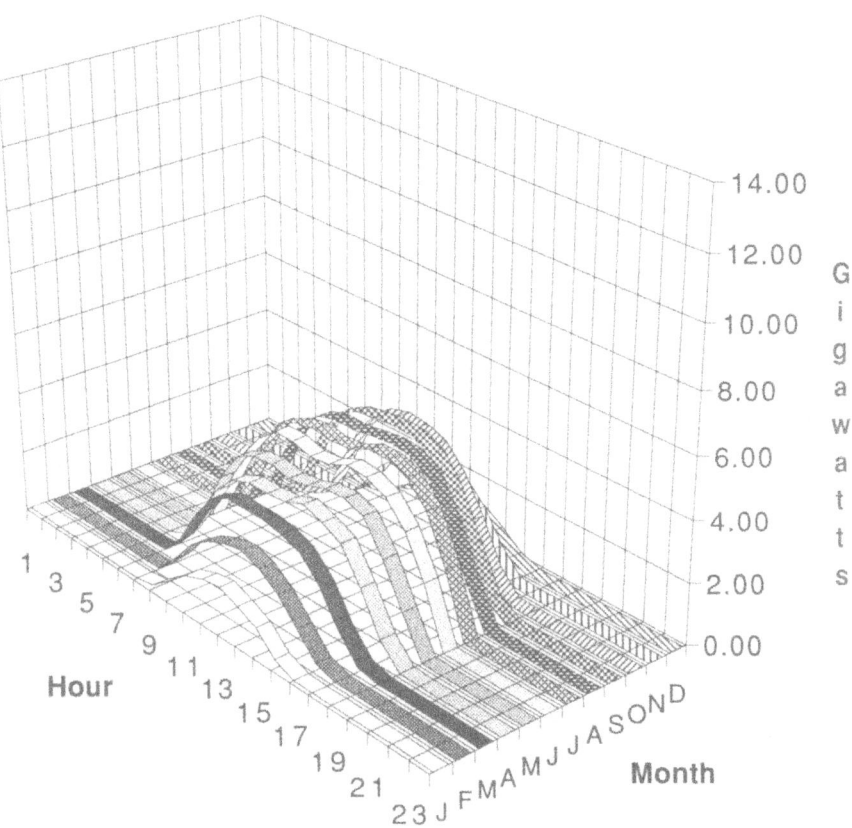

Figure 6.5 Polar collector output pattern as a function of time of day and season, showing the maximum allowed input into the grid before peak lopping intrudes into baseload; the solar fraction is about 12% of total load.
Source: Mills and Keepin 1993.

concluded that this volume is sufficient to achieve 80–90% of total plant output from solar, even where the collector's seasonal output only crudely matches annual load. Backup reduces drastically the amount of storage required because extended periods of cloud are accommodated by the backup fuel, while use of storage also dramatically decreases backup fuel requirements. Modelling runs were done where solar supply amounted to up to 92.5% of the total grid load. These are not shown here, because they are essentially similar in presentation to the load pattern in Figure 6.4. The remaining 7.5% of the load would be supplied by natural gas or other backup source. If gas superheating were used instead of solar superheating, then the solar fraction will decrease to about 82%.

Could such a system be an improvement operationally as well as environmentally? Current nuclear and fossil baseload plants are inflexible and require the addition of more costly peaking plants to follow the utility load. With the solar thermal system, using gas backup and storage, a nearly perfect match is possible with the entire utility load pattern. This *load-following* solar plant would offer an operational contribution to utility grid systems comparable or superior to that provided by conventional, inflexible baseload plants. If a utility were to utilise many such plants, grid stability would be extremely high, and such a system could be operationally superior to current baseload technology.

Surprisingly, energy from a 'load following' solar plant could conceivably be cheaper than from a peaking plant, because storage allows the use of a smaller 'power block' (the power conversion turbines and ancillary equipment) per unit of collection area. For the load-following plant in Longreach, a 70% capacity factor storage plant would deliver busbar costs (basic generation costs before distribution costs, profit and tax are added) of about $A0.060–0.065 per kWh, with the cost at Wagga about $A0.005 per kWh higher because of increased use of natural gas.

Early markets

In Australia, the cost of conventionally generated electricity is relatively low, due in part to both economic and environmental subsidy. This is a serious obstacle to implementation of solar technologies, as early markets are required for any new technologies to allow establishment of production processes and attainment of efficiencies. The exceptions are Western Australia (WA) and the Northern Territory (NT). Both these are isolated from the eastern national grid, and could play an important role in acquiring STE technology early so that production is built up to levels which can lower costs. Moreover, Australia is fortunate in having an even higher cost market to shoulder early development, these being the remote area power supplies (RAPS) to isolated townships in Western Australia and parts of inland Australia that are currently based on diesel. These have a required capacity in excess of the installed capacity of LUZ technology worldwide when one takes into account capacity factors. Also, several suitable sites in WA and the NT have access to natural gas pipelines. These RAPS offer a prime 'proving' market for STE technologies.

There is a confirmed market for stand-alone power generators with more than 1 MW output totalling about 385 MW in Australia. A further 50–100 MW can be expected to be installed in future mining operations, but capacities have not been reported yet. The largest potential exists in WA, where there is a demand of 89 MW in small town supplies, and an additional 215 MW at remote mining sites. This means that Australia could work up to large scale production, without subsidy, within its own borders.

Future grid generation mixes

Because intermittent renewable energy technologies such as wind, wave, and PV have no fossil fuel use component except in their construction, use of these in a generation mix with STE would be possible, with STE providing off-hours storage capability. However, as with the existing power system, the intermittent renewable fraction cannot be made too large without the risk of raising the cost of the STE system. Multiple sources would provide some smoothing of weather fluctuation effects, but the main benefit of including other renewable sources in the mix would be where another renewable technology is cheaper, for example wind energy in a particularly good site. It is doubtful whether flat panel PV will be cheaper than STE for the foreseeable future, but use of PV with inexpensive concentrator technology might be viable after the year 2000.

The absence of cheap electrical storage therefore determines, in large part, the energy generation mix and grid characteristics of the next few decades. Some allowance in this must be made for transition from fossil fuels, and the characteristics of these fuels must also be used to best advantage. In the first decade or so, the current oversupply of baseload coal-fired capacity will have to be wound down, and this will bias new renewable generation to use in two ways. First, STE storage plants can be optimised to provide output within high load periods of the day (say between 7 am and 10 pm). Second, intermittent low cost renewables (wind generation and/or possibly PV concentrators) which

remain at a low enough market share level can be used so that fluc-
tuations in supply can be filled in by these other sources without grid
disruption.

Later, as baseload capacity has to be replaced, twenty-four hour STE
plants would be installed, or existing STE plants modified to operate
over a wider duty cycle. The conventional situation of cheap baseload
(coal) energy and more expensive (gas turbine) peaking electrical
energy would be replaced by a moderately priced STE storage technology, supplemented where possible by less expensive intermittent
renewable sources. This will allow CO_2 emissions from electricity generation to meet long term atmospheric stabilisation goals currently
expected for beyond the year 2020. This would be the major stage of
abandonment of fossil generation, occurring between 2005 and 2020.

Finally, development of renewable chemical fuels or biofuels for
backup of STE plants, and inexpensive electrical storage for electrical
output renewable sources would allow a sustainable, zero-emission
electricity generation system. This does not appear to be necessary for
atmospheric stabilisation at present, but may be necessary as we
become more knowledgeable and strict about the effects of human
activity on the environment and require further reductions in CO_2
emissions.

Solar heat

Apart from the conversion of solar thermal energy into electricity, other
potentially valuable applications of solar thermal energy exist. These
are industrial and domestic applications of solar heat.

The use of energy in the form of heat is very important in modern
societies, with about three-quarters of domestic and most industrial
final energy use being thermal (Wilkenfeld and Energetics 1991). Heat
from renewable energy conversion equipment is not easily transportable, but for many applications on-site solar collector equipment
could be developed to carry much of this load. Current low temperature collectors use 'flat plates' which are very good absorbers of
solar energy but have high surface areas for convective and radiative
loss. At temperatures below about 70°C, such as for residential hot
water, the losses are moderate and the technology is efficient, but at
higher temperatures such losses become serious. The cost of this
technology is approximately the same as for gas or off-peak electricity
when amortised over the lifetime of the device, but marketing and
financing arrangements which allow such amortisation are rarely
available to the solar consumer (see discussion of solar hot water in
Chapter 7).

Solar industrial heat

Since most industrial applications are above 100°C, flat-plate collectors are unsuitable for major industrial markets. However, for higher temperature industrial applications, cost-effective trough and dish technologies will be available during this decade.

A solar industrial heat system developed at the University of Sydney, using seasonally adjusted parabolic trough concentrators, produces steam at temperatures up to about 200°C (Mills 1991a). A demonstration plant has recently been installed at Liverpool Hospital in Sydney, using evacuated absorber tubes. (An earlier version of this system was previously installed on the roof of Campbelltown Hospital.) Preliminary calculations indicate that, when mass-produced, the system is likely to become competitive with process from LPG and so the initial market is likely to be in rural areas. Later, when a larger scale of production is achieved, more advanced solar heat systems could possibly become competitive with natural gas. The main non-technical barriers are the inadequate funding of commercial development in Australia and inadequate action by our governments to help create the initial market through their procurement policies. These two necessary actions must be actively coordinated rather than left to chance.

Markets for applications up to 400°C are potentially accessible to advanced evacuated tube technology using non-focusing or parabolic reflectors to collect solar energy and direct it to the tube. Evacuated tubes prevent convection losses by surrounding a tubular absorber with a vacuum contained within an outer transparent glass tube. They suppress re-radiation of heat with special spectrally selective absorbing surfaces, which absorb well in the solar spectrum, but radiate poorly in the infrared re-radiation spectrum. Commercial tubes currently available are suitable for efficient operation up to 200°C, and advanced selective surfaces suitable for operation up to much higher temperatures are under development. These technologies will be available at much the same lifetime-amortised cost as conventional fuels. Use of these technologies, backed up by direct natural gas, is one of the most effective pollution and greenhouse gas emission reduction opportunities possible. They can achieve reductions in CO_2 emissions of up to 90%, and can be thus regarded as close to energy efficiency in terms of environmental effectiveness.

Markets above 400°C are also potentially accessible by versions of point-focus and central receiver technologies. Higher temperatures than are possible for parabolic troughs are possible for collectors such as would be used in some solar thermal electric plants, for example, paraboloidal dishes with accurate tracking (Hagen and Kaneff 1991). Such technology was discussed in the previous section.

Solar residential heat

Because of the slow turnover of housing stock, it will be some time before passive solar heating can make its full contribution to the reduction of pollution and greenhouse gas emissions. However, a low-cost solar space heating system, easily installed on existing homes, would have a very large potential market and could make a rapid contribution.

A fixed version of the University of Sydney solar steam collector system is being successfully demonstrated on a house in Sydney, funded by Pacific Power (see Kentish 1993). In the domestic system, the tilt of the rooftop collectors does not have to be adjusted after installation. Rather, the shape and tilt of the collectors is optimised for winter at the latitude of installation. Energy is stored in a tank in the form of hot water. The domestic system provides both space heat and hot water; a separate unit also provides the heat for a cooking stove. Because of the inbuilt storage of the system, the electrical backup can be supplied at off-peak periods. In a home which uses electricity for space heating and hot water, the solar heating system and energy efficiency measures could together reduce total fossil fuel use by 75% (Mills 1991b).

Conclusion

To conclude, a summary of the situation with solar thermal energy can be given. Several technologies can be developed to provide relatively inexpensive solar heat and solar thermal electric power. Domestic solar thermal energy is a reality but needs expansion and improved financing arrangements to realise full potential. Thermal energy for industry up to 350°C will soon be available at attractive prices.

For the single, most important sector, electricity production, the following can be shown:

- The optimal economic and environmental configuration for producing solar thermal electricity will be a plant that has both storage and backup fuel.
- Load-following solar thermal plants with storage offer lower CO_2 emissions than mixes of non-storage renewables (PV, wind) backed up by fossil plant. Overall, carbon emissions are far lower than with advanced or improved fossil fuel scenarios.
- Inexpensive solar thermal storage such as rock beds can provide the storage component.
- Storage plants have operational and design flexibility – a plant which is installed as a peakload plant with limited storage can be easily turned into a baseload plant by adding more collector area and

storage units. Thus early installations, which may be more attractive as peaking plants, can be adapted as the need for solar baseload grows. Solar repowering of conventional plants is also feasible.

- Preliminary economic estimates show that suggested design improvements could result in extensive solar thermal electricity at a price of $A0.06–0.07 per kWh over large areas of Australia. Such price and availability could make solar thermal technology highly competitive for new electricity supply, possibly before the year 2000.
- Off-grid markets in Australia could provide an appropriate, useful and lucrative initial market for these technologies.
- Buyback prices (paid to generators for contribution to the main grid) are becoming increasingly attractive, particularly in Western Australia, but high discount rates could erode this advantage. Adequate accounting for the environmental externalities of fossil fuel electricity would raise buyback rates considerably.
- Storage solar thermal plants may be cheaper than peaking solar thermal plants and the difference between the costs of energy from hybrid storage plants in high solar radiation regions (e.g. Longreach) and poorer ones (e.g. Wagga Wagga) is only $A0.005 per kWh. Thus, the technology can be installed widely over the Australian mainland.

This chapter has made it clear that solar thermal technologies possess considerable potential in the near term. If the criteria for energy reform is to reduce our reliance on fossil fuels, and thus our greenhouse gas emissions, then solar thermal electricity, possibly in partnership with fossil fuels or other renewable technologies such as wind power (see Chapter 7), offers the largest single area of potential gain in energy supply technology.

References

Australian Bureau of Meteorology. 1975. *Climatic averages, Australia.* Canberra: Australian Government Publishing Service.

Blakers, A., Crawford, T., Diesendorf, M., Hill, G. and Outhred, O. 1991. *Opportunities for the Australian wind energy industry in reducing greenhouse gas emissions.* Report to the Department of the Arts, Sport, the Environment, Tourism and Territories. Sydney: UNISEARCH.

Dayton, L. 1991. Ceramic coat makes the most of sunlight. *New Scientist.* 30 November: 21.

Dubberley, L.J., Gormley, J., Lang, W., Liffengren, D., McKenzie, A. and Porter, R. 1981a. *Comparative ranking solar thermal systems: cost and performance of thermal storage concepts in solar thermal systems, phase 1, volume 2.* SERI/TR-631-1283. Golden, Colorado: Solar Energy Research Institute.

Dubberley, L.J., Gormley, J., Lang, W., Liffengren, D., McKenzie, A. and Porter,

R. 1981b. *Cost and performance of thermal storage concepts in solar thermal systems, phase 2.* SERI/TR-XP-0-9001-1-B. Golden, Colorado: Solar Energy Research Institute.

Hagen, D.L. and Kaneff, S. 1991. *Application of solar thermal technologies in reducing greenhouse gas emissions.* Report to the Commonwealth Department of the Arts, Sport, the Environment, Tourism and Territories. Canberra: ANUTECH Pty Ltd.

Jaffe, D., Friedlander, S. and Kearney, D. 1987. The LUZ solar generating systems in California. In: *Advances in solar energy*, p. 519, Proceedings, International Solar Energy Society Congress, Hamburg, Germany, September 1987. Workshop 2.W3.

Kentish, J. 1993. On the sunny side of the street. *Australian Geographic.* 30: 20–21.

Mills, D. 1991a. Atmospheric stabilisation approach to the industrial sector: solar process steam and solar thermal electric generation. Paper to the Ecologically Sustainable Development Workshop on *Energy Efficiency and Renewable Energy*, Canberra, 16–18 April 1991.

Mills, D. 1991b. Atmospheric stabilisation approach for the domestic sector: a retrograde integrated solar/efficiency approach. Paper to the Ecologically Sustainable Development Workshop on *Energy Efficiency and Renewable Energy*, Canberra, 16–18 April 1991.

Mills, D.R. and Keepin, W. 1993. Baseload solar power. *Energy Policy.* August 1993: 841–858.

Mills, D.R. and Morrison, G.L. 1991. *Solid energy storage system for industrial heat applications.* Engineering feasibility study for the Local Government Electricity Association of New South Wales. Department of Applied Physics, University of Sydney.

Solar Energy Research Institute. 1988. *Phase change thermal energy storage: report on symposium, Helendale, California, October 19–20 1988.* SERI/STR-250-3516. Golden, Colorado: SERI.

Swet, C.J. 1990. Energy storage for solar systems. In: De Winter, F. (ed.), *Solar collectors, energy storage, and materials.* Chapters 14–19. Cambridge MA: MIT Press.

Trombe, F. and Le Phat Vinh, A. 1973. Thousand kW solar furnace built by the National Centre of Scientific Research in Odeillo (France). *Solar Energy.* 15: 57.

Vant-Hull, L.L. 1991. Solar thermal electricity: an environmentally benign and viable alternative. In: *Proceedings, World Clean Energy Conference*, Geneva, November 1991. pp. 350–356. Zurich: Cercle Mondial Du Consensus.

Wilkenfeld, G. and Energetics Ltd. 1991. *Greenhouse gas emissions from the Australian energy system.* National Energy Research, Development and Demonstration Corporation end of grant report 1379. Canberra: Department of Primary Industries and Energy.

Zhang, Q.C. and Mills, D.R. 1992. Very low emittance solar selective surface using new film structures. *Journal of Applied Physics.* 72(7): 3013–3021.

CHAPTER 7

Wind, biomass and other renewables

MARK DIESENDORF

This chapter surveys wind, biomass and other renewable energy sources. It examines the current state of technological and economic viability, and the policy context for adoption. More space is devoted to those technologies that are more suitable for implementation or expansion in the near term.

Windpower

Wind energy has been harnessed for human use for thousands of years. Sailing ships, windpumps for lifting water and windmills for grinding grain are all mentioned in ancient writings from several centuries BC. Electricity generation from the wind appeared in Europe at the end of the 19th century. In the 1940s, *Dunlite* wind generators for charging batteries were a familiar sight in rural Australia, but the rapid spread of electricity grids in the 1950s and 1960s, assisted by subsidised grid connection fees and cross-subsidised rural electricity rates, fore-shadowed the end of this indigenous product.

In the 1970s, the sudden large increases in oil prices, together with the growing public awareness of the environmental and health hazards of fossil fuels and nuclear power, stimulated a revival in wind electric power in Denmark. By investing in a national wind energy site survey, a wind generator test station, research and development on wind technology, and subsidies to purchasers of wind generators, the Danish Government created an expanding domestic market for windpower. Local farm machinery companies manufactured the first wind generators.

In the USA in the early 1980s, tax incentives were given for installing renewable energy and legislation was passed requiring utilities to pay a fair price for electricity sold to the grid by private generators. This gave windpower the opportunity to compete with the heavily subsidised fossil fuel and nuclear industries. In California especially, government incentives and regulatory changes stimulated the formation of a large market for wind power, just when the domestic windpower industry in Denmark had become strong enough to expand into exports.

By 1985, windpower had became a major export industry for Denmark. Export and domestic sales together had created, directly, over three-thousand new person-years of employment. The Danish Government's initial investments in wind energy had already been paid back two to four times over, just in terms of increased income tax revenue and reduced unemployment payments (Møller 1985). Through the late 1980s, government subsidies were gradually phased out.

Current status

In late 1992, there were about 2500 megawatts (MW) of wind plant installed globally. The vast majority of this plant comprises wind generators rated in the range from 50 to 400 kilowatts (kW) each and connected to large electricity grids. Approximately 1600 MW is located in California where it supplies about 2% of annual electricity generation. Most of the remainder is in Europe. In Denmark, about 450 MW windpower provides about 3% of electricity and this is expanding toward the government's target of 1500 MW, about 10% of the electricity generated annually, by 2005 (Danish Ministry of Energy 1990). Other rapidly growing windpower programs are in Germany, the Netherlands and India.

Australia's first tiny wind farm (array of wind generators), comprising six 60 kW wind generators manufactured by *Westwind Turbines* of Perth, is located at the isolated town of Esperance in southern WA, where it operates as a fuel saver for the town's diesel-powered generation plant. Further installations at Esperance lifted wind capacity there to 2.4 MW in early 1994. There are also diverse single wind generators of capacity greater than 30 kW scattered around the country, as shown in Table 7.1. Unfortunately, there is not a single medium-sized wind generator connected to the state grid in two of the states with the largest wind energy potential, Tasmania and South Australia. A 10-MW wind farm proposed for Victoria might be the first wind farm to be connected to an Australian state electricity grid.

Technology

Through the 1980s there were improvements in technology which increased the efficiency of wind energy conversion and the reliability of wind generators, and hence reduced the capital and operating costs of

Table 7.1 *Wind generators of capacity greater than 30 kW operating in Australia in late 1993*

Company	Capacity (kW)	Country of origin	Location	Date of installation
Nordtank	55	Denmark	Rottnest Is., WA	1983
Westwind	60	part Australia (imported blades, alternator, etc.)	South Fremantle, WA	1984
	60		Woodman Point, WA	1985
	6 × 60		Esperance, WA	1986
	60		Breamlea, Vic.	1987
Windmaster	175	Belgium/Holland	Malabar, Sydney, NSW	1985
Foden	55	part Australia	Flinders Is., Tas.	unknown
Nordex	150	Denmark	Coober Pedy, SA	1991
Vestas	9 × 225	Denmark	Esperance, WA	1993

windpower substantially. The economically optimum size of individual wind generators increased from 50 kW in the early 1980s to 250-400 kW by 1991. Today the best machines are available to generate power for over 95% of the time and, at sufficiently windy sites, have capacity factors (annual average power divided by peak power) of over 30% (Blakers *et al.* 1991a).

The most common type of modern wind generator has a turbine consisting of two or three blades, with an aerofoil cross-section, rotating at fixed speed around a horizontal axis. The turbine drives a generator which produces alternating current at fixed frequency. Further techno-logical improvements which can be expected to enter the market over the next five years include the following:

- development of aerofoil designs which are more suitable for wind turbine blades than those based on the requirements of aircraft wings or helicopter blades;
- allowing the blades to rotate at variable speeds increases the effici-ency of energy conversion. Modern power electronics then produces fixed frequency alternating current (a US company currently market-ing a 300 kW wind generator of this type which is claimed to supply electricity at $US0.05/kWh at favourable sites);
- the use of high strength composite materials currently being devel-oped for aircraft might allow the optimum size of wind generators to be increased; and
- the manufacture of wind generators which could be easily optimised in performance to the wind conditions of particular sites, after installation.

Such developments would be expected to improve the viability and efficiency of windpower significantly in the near future.

Siting

The choice of sites for wind generators is crucial to the economics of wind generation. This is because the power in the wind is proportional to the cube of the windspeed. In other words, if the windspeed is doubled, the windpower is increased eight times. The windiest sites tend to be found on exposed, treeless areas of coastline, hills, ridges, and mountain passes. Cities and suburbs are generally unsuitable, because the presence of buildings even at a low density slows the wind substantially. Globally, there is known to be substantial windpower potential in northern Europe (including Scotland and western Ireland), large areas of the USA, the south island of New Zealand and the southern coastline of Australia (including much of Tasmania's coastline). To some extent, windpower potential tends to be higher in areas where the potential for the direct use of solar energy is lower, as illustrated for Australia in Figure 6.1 in the previous chapter.

The windpower potential of a particular area is constrained by the other land uses allocated to that area (e.g. housing, airport approach, national park) and also depends on the costs of other sources of power in that area. For example, in the south-west of Western Australia, grid-connected windpower may only be competitive with coal-fired power at sites where the annual mean windspeed is at least 8 metres/second, but in parts of California windpower is competitive with conventional power sources provided the annual mean windspeed exceeds 6.5 m/s. This is because conventional power is more expensive in California than in Australia.

Denmark has recently installed the first offshore demonstration wind farm in the Baltic. In densely populated countries, it is hoped that the higher windspeeds available offshore will compensate for the additional costs of construction in shallow waters and transmission to land.

Environmental impact

During operation, wind plant emits no chemical pollutants. Per unit of electricity generated, the materials required to manufacture wind plant are similar in quantity and pollution production to those of a corresponding coal-fired power station. However, the pollution arising from the construction of a wind or coal-fired power plant is dwarfed by the pollution from the operation of a coal-fired power station. So, a shift from coal to wind would lead to a dramatic reduction in carbon emissions and other pollution.

In Europe, wind generators are not regarded as a hazard to birds, but there has been one site in the USA which experienced special problems in this regard, requiring modification to electrical conductors. Although wind generators do emit noise, this has not been a widespread problem because they are not normally sited close to residential areas. In Europe, noise from wind generators is controlled by general noise standards. In the densely populated Netherlands, the government also gives bonuses to manufacturers of quiet machines, and this seems to be reducing noise levels. Wind generators can coexist with most types of farming. The land occupied by wind generators and their access roads is about 1% of the land area spanned, and is generally smaller than that occupied by an equivalent coal-fired power station and associated open-cut coal mines.

The most common objection to wind generators is based on their visual impact. All human-built structures have some effect in this regard. This is a highly subjective area, involving a personal judgement as to the relative aesthetic merits of a wind farm versus such alternatives as a coal mine and power station. (Also, arguably the worst environmental impact of a coal-fired power station is its emission of greenhouse gases, which are invisible.) Public surveys suggest that visual impact of wind generators can be reduced by sensitive siting (e.g. avoidance of national parks, World Heritage Areas and old growth forests). Another means of reducing visual impact is through 'visual unity'; that is, at a particular site, ensuring that all wind generators have the same number of blades, the same direction of rotation and the same type of tower. There is some evidence that, once wind farms have been installed for several years, they tend to become increasingly regarded as part of the landscape, just like church steeples in European villages.

Economics

Another question often raised about windpower is that of relative cost. There are economies of scale in manufacturing, installing and maintaining wind plant. The cost of wind electricity decreases as the size of the wind farm increases up to 50–100 MW capacity. For an installation of the latter size at a very good site in the country of manufacture, the cost of electricity in Denmark or California in 1992 can be as low as $US0.05/kWh. This is a 'levelised annuity' costing, which is based on a discount rate of 7% and a plant lifetime of twenty years. Naturally the costs would be higher if the wind plant is exported to another country and/or if the site is remote. The technological improvements described above will further reduce costs.

In the early days of windpower, some utility engineers claimed incorrectly that windpower only has economic value as a fuel saver in conventional power plant. They argued that wind plant could not substitute for the capital costs of conventional power plant, which would always have to be kept in reserve for the times when the wind does not blow. The fallacy in this argument is that it assumes that conventional power plant is completely reliable while wind plant is totally unreliable. In reality, both types of power plant are partially reliable. Even in a conventional electricity grid, there is reserve plant amounting to 15–25% of the total installed capacity of the grid. When windpower is installed in the grid, this reserve plant can also serve as most of the backup for the wind plant, provided wind generates a small fraction of the total power output of the electricity grid. To keep the overall reliability of the grid at its previous level, some additional gas turbines may have to be installed, but these have low capital cost and are rarely operated and so may be considered to be low-premium reliability insurance for the wind plant (Blakers *et al.* 1991a).

There are other economic values of wind plant which are difficult t(
price, but none the less real. They relate to environmental benefit£
small module size and short construction and installation times. Unde
uncertainty about future power demand, the latter two properties mea1
that windpower (and indeed most renewable sources of energy) hav(
less financial risk than large power stations.

Barriers to windpower

Notwithstanding the potential improvements in windpower technolog
being pursued, the main obstacles to using windpower are not technic£
ones. In Australia, barriers to the more rapid deployment of windpowe
include:

- the high discount rates (15–20%) faced by private generators con
pared with the low discount rates (7–8%) faced by utilities;
- the unfairly low buyback prices offered by utilities for the sale of ele
tricity to the grid (these prices fail to account for the partial reliabilit
environmental benefits and reduced financial risk of wind plant);
- the political influence of the coal industry, which may have resulted i
the cancellation of a proposed 20 MW grid-connected wind farm i
Western Australia and the almost simultaneous announcement of
coal-fired power station in a marginal electorate of that State (:
Western Australia, the unit cost of electricity from such a station cou
be comparable with that of a large wind farm (Diesendorf 1993));
- the proposed strengthening of the eastern and south Australi£
electricity grid, which would enable cheap electricity from the exce
capacity of coal-fired powered stations in New South Wales ar
Victoria to be sold to Tasmania and South Australia, two States wi
large wind potential; and
- the virtual absence of research, development and demonstratic
funding for wind energy in Australia. Indeed, it appears that m(
current research is being conducted by critics of windpower.

The question now is how these barriers can be overcome. Lesso
can be drawn from current actions overseas. New Danish legislati(
specifies the buyback price for wind generated electricity at 85% of t
consumer price of electricity. Where windpower is fed back into t
grid, the utility is responsible for strengthening the grid, where nec
sary, while the wind generator owner pays for the local connection
the 10 or 20 kilovolt distribution line.

In the USA, the Federal Energy Bill passed in 1992 provides a $1
0.015/kWh tax credit or, for companies not paying tax, a producti
payment, for wind-generated electricity. The payment recognises,

part, the environmental benefits of windpower. It applies to new wind farms whose construction commences in the period from the beginning of 1994 to mid-1999. The payment will continue for ten years and will be adjusted for inflation. The Bill also ensures transmission access for windpower at prices to be determined by the Federal Energy Regulatory Commission. Unlike Australia, both the USA and the European Community have substantial budgets for windpower research and development.

Energy from biomass

Biomass energy sources include wood, crops, crop residues and manure. Biomass can be burnt to produce useful heat and/or electricity, or converted into liquid or gaseous fuels. Provided the biomass which is used in this way is replanted, the combustion of biomass or biomass fuels produces no net increase in greenhouse gas emissions. The CO_2 given off during combustion is, in general, balanced by the CO_2 sequestered during growth of the next crop. However, some local pollution is still produced.

Several liquid fuels obtained from biomass can be used as substitutes for petroleum-based fuels. These include:

- ethanol produced in the traditional way by fermentation and distillation of the sugar or starch content of dedicated crops (such as sugar cane, sugar beet, fodder beet, corn or wheat) and used as an extender for petrol or diesel;
- methanol produced by gasification and synthesis of the ligno-cellulose in woody crops and wastes;
- vegetable oils as substitutes for diesel; and
- biogas produced by the anaerobic digestion of manure and plant wastes as an alternative to natural gas.

However, large scale centralised production of biofuels by traditional methods could be very expensive and damaging to the environment through nutrient loss from soils and degradation of marginal land. Moreover, some existing methods of production and distribution of liquid fuels from biomass without replanting create almost as much greenhouse gas emissions as the production and use of petroleum fuels. Fortunately, there are at least two approaches to the production of useful energy from biomass which are low in economic cost and environmental impact: the use of biomass wastes and ethanol production using biotechnology.

Energy from biomass wastes

At present the most economical conversion of biomass into useful energy is to burn wastes such as bagasse, a by-product of the cane sugar industry, or wood wastes from the timber industry. Bagasse is currently burnt very inefficiently to provide the energy for sugar refining. Small modern power plants could be installed at sugar refineries to produce electricity for the grid. Although the costs are likely to be competitive with those of power from fossil fuels, the low buyback rates offered by the electricity commissions are a disincentive. In Queensland, the total electricity generation could amount to several hundred megawatts at least (Edwards 1990; Hagen & Kaneff 1991).

Firewood can be an economical, renewable source of energy, although it is currently not being used sustainably. In the USA, firewood provides approximately the same quantity of end-use energy as nuclear energy (but, of course, firewood provides energy mostly in the form of industrial and domestic heat, while nuclear energy provides electricity). In Australia, the contribution of wood to total energy use is not insignificant (see Chapter 2). In Tasmania, much domestic heat is currently provided by firewood which is waste from the forestry industry. But, for ecologically sustainable production, firewood plantations would be required on land which has already been cleared. Some species of trees in plantations can provide five to ten times the amount of fuel per hectare as native forests. But unless the plantation has more than one economic function (e.g. land restoration as well as firewood), the cost of energy from plantation firewood could be considerably higher than from wood waste.

For efficient end-use, open fires must be replaced with efficient, airtight slow combustion stoves, especially in urban areas. These stoves use less wood, and can produce less pollution. There are currently several Australian-made stoves of varying degrees of efficiency on the market. Research on the efficient burning of firewood has been done by Dr John Todd at the University of Tasmania. More research is needed on the quantity of chimney emissions and their health hazards, but unfortunately there is at present negligible funding for this work in Australia (for further discussion, see Todd and Singline 1989).

A limited amount of liquid fuels can also be produced from biomass wastes. For example, the Manildra Group, the largest wheat flour miller in New South Wales, has recently installed a fermentation and distillation plant to convert waste starch into ethanol. In the course of the company's production of gluten, the waste starch was part of an effluent stream which previously was used to fertilise land for growing grass for grazing cattle. Now ethanol is produced from the waste stream and this is mixed with diesel to form diesohol E15, comprising 15% ethanol, 84.5% diesel and 0.5% emulsifier. It is possible that up to 30% by volume of ethanol could be used in existing diesel engines with little or no modification (Reeves and Lom 1992). The diesohol is being sold at the same price as diesel to run several fleets of trucks in Sydney and several buses in Canberra. In addition to reducing the greenhouse gas emissions from diesel by up to 15%, the use of diesohol E15 also reduces local pollution from carbon monoxide, nitrogen oxides and hydrocarbons.

Currently, ethanol does not attract fuel excise and under this condition the Manildra project is economically viable. Production could be expanded considerably if ethanol were granted a long term exemption from excise.

Ethanol production by means of biotechnology

Even with such an exemption, ethanol production from biomass wastes will still be limited by the size of the market for the principal product (e.g. gluten in the case of the Manildra Group). In Australia it would be too expensive to grow wheat just for ethanol production.

However, recent work by APACE Research Ltd and the Biotechnology Department at the University of New South Wales offers a potentially low-cost method, which is also low in environmental impacts, for producing large quantities of ethanol from dedicated crops as well as from wastes. The new method, currently a laboratory process, involves the use of biotechnology to produce ethanol from lignocellulose (the wood or fibre content of biomass). The method has a much higher energy efficiency than traditional ethanol production, and simultaneously allows further energy to be extracted from a solid by-product of this process, lignin, which could be used as a fuel for steam generation. This new method of producing ethanol can be applied both to the traditional crops such as sugar cane, cassava, wheat, corn and barley, and also to dedicated lignocellulosic crops, such as sweet sorghum, kenaf and trees (Reeves and Lom 1992).

Because it is expensive in dollars and energy to transport biomass over long distances and because lignocellulosic feedstocks are highly decentralised and dispersed, the development of such a biofuels industry would need to be based in regional rural centres. If based on tree plantations, it could simultaneously assist land restoration projects and encourage regional industrial development by producing fuel. In this way, environmentally beneficial employment could be created in rural areas. Such potential new local industries are currently disadvantaged by inadequate funding for research, development and demonstration projects.

Solar photovoltaic cells

Solar photovoltaic (PV) cells convert sunlight directly into electricity. Since the early 1980s, steady advances have been made in the technology, with the result that their cost has been decreasing and their performance has been improving (Blakers *et al.* 1991b). The current cost of bulk orders of PV modules is A$6000–7000 per peak kilowatt installed, which is about one-third of the cost in the late 1970s, after correcting for inflation. Worldwide, about 200 MW has been installed, with about 50 MW of this installed in 1991. In Australia, these reliable systems are used for water pumping and charging batteries to power small homes, signals and telecommunication repeaters which are isolated from the grid. For isolated homesteads, there is an immediate additional market for about 20 MW of PV modules as components of Remote Area Power Supply (RAPS) systems, consisting of PVs, batteries, inverter (to change direct into alternating current) and diesel generator. In a RAPS system the cost of the PV modules and batteries is offset by the greatly reduced fuel cost, the lower capital cost of the diesel (because a smaller diesel unit can be purchased) and the considerably reduced maintenance cost of the diesel (because it can be run at full power for short periods instead of at low power for long periods).

The potential market for RAPS is being undermined in several Australian states by uneconomic extensions of electricity grids into rural areas, subsidies for the connection of homes and farms to the grid, and by State Government policies to charge uniform electricity tariffs over their states. Although electricity utilities do provide diesel generators for isolated communities in some states and the Northern Territory, they do not normally include PVs and batteries in these systems, despite their economic competitiveness for this purpose.

Photovoltaics are still much too expensive for the production of electricity for the grid, although this may not be the case within ten to fifteen years. Their advantages over wind power and solar thermal electricity are that their modules are small and lightweight, and so are easily installed on the rooftops of homes and places close to the point of use. They have no moving parts or working fluids, and so the system maintenance costs are very low, provided the batteries are properly looked after.

Two companies, BP Solar and Solarex, manufacture solar cells and modules in Australia. At present their combined output is 1.5 to 2 MW per year, and about 40% of this is exported. Although the Australian market is constrained by the non-technical and non-economic barriers mentioned above, there are large and growing export markets in Asia. Today's solar cells for electric power production are made from silicon crystal wafers. Over the next five to ten years major reductions in costs could come from using thin films, which are cheaper but less efficient than silicon crystal wafers; and using sunlight concentrators with expensive but highly efficient solar cells.

Passive solar housing

The use of the sun to warm houses in winter, and conversely designing houses to keep the sun out and thus be cooler in summer, is perhaps the simplest of all renewable energy options, and directly replaces the need for use of fossil fuels and electricity. Energy efficient, passive solar houses are now being built by public housing authorities and several private builders, where such design can reduce the cost of winter heating by 60–90% and can also make homes much more comfortable in summer (Ballinger *et al.* 1992). The techniques available include dwelling siting for good orientation, insulation, draught-proofing, and improved window placement and design.

More rapid proliferation of passive solar housing is being inhibited by the failure of local government and the housing industry to orientate streets and subdivide blocks in an appropriate manner, and by the slowness of Federal, State and Territory Governments in developing and implementing energy performance standards for new buildings. Victoria has introduced compulsory insulation for some types of new homes, but this is not the same as energy performance standards, which would offer architects, builders and/or future home owners more flexibility in choosing the means to achieve the standards.

Considerable energy savings are possible via passive solar housing, both in designing new dwellings and in retrofitting existing ones. Importantly, a well-understood framework for reform already exists in the form of building regulations and ordinances. This is clearly an area where the only real barrier is the lack of political and community will to implement the changes.

Solar hot water

Solar hot water, based on flat-plate collectors, is widespread in the Northern Territory and, to a lesser extent, in Western Australia. It is also used in other parts of Australia. It is a well developed and proven technology, and can contribute significantly to reducing fossil fuel use and associated waste production. However, its further expansion is being held back over much of the continent by the absurdly low prices charged for off-peak electric hot water produced by coal-fired power stations (see Versluis 1991). In several Australian states these prices are artificially low because of excess capacity in their electricity grids.

Solar hot water is also disadvantaged by inequitable financing. The full, lifetime costs of a solar system have to be paid up front by the consumer when the hot water system is purchased, but the power to run the system is henceforth free. In contrast, the cost of the power station to generate electricity for electric hot water systems is generally financed by means of a government guaranteed load paid off at low interest rates over 20 to 30 years by electricity commissions and trusts. Purchasers of electric hot water systems do not 'see' this cost at the time of purchasing a hot water system, as it is spread out on their electricity bills. The longer term benefits of the solar option, both economic and environmental, remain concealed. These inequities could be removed by the introduction of integrated least-cost energy planning (see ESD Working Groups 1991; Deni Greene Consultancy Services 1991).

Hydro-electricity

During operation, hydro-elecricity produces negligible atmospheric pollutants. This is a strong environmental advantage, but one that has to be weighed against the environmental costs of land inundation and disturbance of rivers. In the generally dry and variable Australian environment, potential for hydro-electricity was never as great as in wetter countries. Most of Australia's large-scale, good potential sites have already been developed. Most hydro power is generated in Tasmania and in the mainland alps (notably the Snowy Mountains Scheme). This Scheme and others also serve to store irrigation water. Those sites which are still unexploited tend to lie in areas of great scenic beauty or ecological value, such as the south-west of Tasmania, where the environment movement successfully stopped the Gordon-below-Franklin scheme in 1983, and the far north of Queensland, where there is currently opposition to the proposed Tully–Millstream scheme. The latter would lie partly in the Wet Tropics World Heritage Area. However, there is still some potential for smaller scale additions or increments to hydro in Australia.

Low-cost schemes with minimal environment impact are possible, simply involving the installation of generating systems on existing water supply storages. This is essentially the use of an untapped resource. There is also some potential for new, small-scale hydro schemes (Williams 1991). Although subject to assessment on a case-by-case basis, these have much less environmental impact than large dams. Finally, very small ('micro-hydro') schemes can be built which may not require dams at all, on some rural properties (Lemon 1991). Two Australian companies are manufacturing and exporting micro-hydro generators: Tamar Designs, located near Launceston, Tasmania, and The Rainbow Power Company, at Nimbin in New South Wales.

Vanadium redox battery

The vanadium-redox battery, which is being developed by Professor Maria Skyllas-Kazacos and colleagues at the University of New South Wales, is very different from conventional ones (see Skyllas-Kazacos 1991). Energy is not stored in the reactor itself, but rather as vanadium liquid in two separate tanks from which the liquid is pumped through the reactor. The energy storage capacity can be increased by enlarging the tanks. There are no standing losses of energy. The efficiency of storage and retrieval of energy is very high: 87% (including pump energy), compared with about 70% for lead–acid batteries. Vanadium is a common element in the Earth's crust, and so its price could be quite low once the batteries are being mass-produced and sold on a large scale. Moreover, spent vanadium liquid could be recharged by means of a renewable energy source and used again and again.

In comparison with existing batteries, the vanadium redox has low energy density. This means that, to power an electric vehicle over a reasonable range, the tanks would have to be much larger than the equivalent petrol tank. This disadvantage could be reduced by more-frequent 'refuelling', bearing in mind that this would simply involve changing the vanadium liquid in the tanks (the spent liquid then being recharged for a subsequent customer at the filling station). 'Refuelling' could be almost as fast as filling up a car with petrol. But there are currently several other types of advanced battery being developed in Australia and overseas (Schodde 1990), and it is not clear whether the vanadium redox battery will prove to be the best option for the large potential market for batteries in electric vehicles.

Because of its low energy density and the absence of a network of 'vanadium stations' for recharging and 'refuelling', the first use of the vanadium redox battery is likely to be as a means of storage for electricity in remote area power supplies (RAPS). For RAPS it could rapidly displace lead–acid batteries, provided the mass production cost estimates of the University of New South Wales group are borne out in practice. In addition to the domestic market in Australia, there would be significant potential for exports to developing countries. The main barriers are the costs of commercial development and of providing the widespread rural infrastructure to support a significant initial market in RAPS. Australian companies have been unwilling to invest in this new technology, and State and Territory governments have not provided the market, although they could have. At present, it looks as if this new Australian technology may be first manufactured under licence in Thailand and eventually imported into Australia.

Electric vehicles

The forthcoming enaction of 'zero emission' legislation to control local pollution from motor vehicles in Los Angeles is creating a market for mass-produced electric vehicles, whose batteries would be charged from the electricity grid. Mass production of such vehicles is expected soon in the USA, Japan and Europe. In California, a large fraction of grid electricity is already supplied by renewable sources: hydro, geo-thermal, wind and solar thermal. However, in regions such as eastern mainland Australia where most electricity is supplied from coal, the benefits of grid-charged electric vehicles in reducing CO_2 emissions are negligible or even negative – the emissions are simply transferred from exhaust pipes to power stations. The improved efficiency of the electric motor over the petrol or diesel engine is often slightly outweighed by the increased CO_2 emissions from coal over oil. In these regions, the following types of vehicles would be preferable to electric vehicles recharged from the grid:

- motor vehicles fuelled by compressed natural gas (CNG) or liquefied petroleum gas (LPG);
- motor vehicles fuelled by mixtures of ethanol with petrol or diesel;
- electric vehicles powered by fuel cells which are fuelled by natural gas or, eventually, hydrogen from renewable energy; and
- most importantly, improved public transport and increased rail freight transport between major centres, to reduce the use of motor vehicles (see Chapter 4).

However, in regions where a significant part of grid electricity comes from renewable sources, electric vehicles can reduce both local pollu-tion and greenhouse gas emissions. Tasmania, where almost all elec-tricity comes from hydro and where there is substantial wind energy potential, is the best place in Australia to encourage the widespread use of electric vehicles.

It has been suggested that, when the costs of solar photovoltaic cells decline sufficiently, they could be installed on garage roofs in sunny towns for the specific purpose of charging electric vehicles. But then, for commuter vehicles using lead–acid batteries, two sets of batteries would be required: one for the vehicle and one for daytime charging in the garage. It may be better to change batteries at a service station each time the vehicle runs low on charge. A vehicle would only require a single vanadium redox battery and, as noted above, this could be 'refuelled' rapidly at service stations and recharged there at leisure from solar energy.

There is insufficient roof area on a typical electric car for the installation of enough solar cells to provide an adequate transport distance, although they could be used in this way to extend the vehicle's range. Weighing against this is a growing awareness in the community that the provision of vast areas of ground level car parking is an economically wasteful and environmentally and socially damaging use of urban land. If personal vehicles are to continue to be used for urban commuting, they would surely be stacked in parking stations.

Hydrogen storage

Ultimately, when renewable energy can be stored and transported on a large scale, it could supply all non-wasteful energy needs: for heating, electricity and transport. The production of energy-rich, transportable fuels to replace oil or gas is a more difficult task than, say, replacing coal-fired electricity with solar thermal. One promising means for the future large-scale and long-term storage and transmission of energy is in the form of hydrogen and various compounds of hydrogen (Gray 1991; Bockris *et al.* 1991). The significant prospect is the production of hydrogen fuels using renewable energy sources; hydrogen can be produced by using wind- or solar-generated electricity to split water (H_2O). Like natural gas, hydrogen can be burnt to produce useful heat or motive power, or can be used to produce electricity in a fuel cell.

When used in these ways, hydrogen is converted back into water. In this water cycle, there is negligible emission of greenhouse gases other than water vapour itself. Although there is no *net* production of water, there is conversion of liquid water into water *vapour* which is emitted into the atmosphere, and this could increase global warming slightly as water vapour is a greenhouse gas. However, water vapour is a short-lived gas, and the quantities of water vapour thus produced would be small in relation to the total in the atmosphere compared to, for example, carbon dioxide produced from fossil fuels. This is an area that requires quantification. The levels of local pollutants, such as oxides of nitrogen, produced by using hydrogen are much less than those produced by a petrol or diesel engine.

Hydrogen is at present much more expensive than its principal competitor, natural gas. This is partly due to significant inefficiencies and costs in the production and storage of hydrogen. Furthermore, the technology and infrastructure are not presently available to use hydrogen on a large scale. If hydrogen were to replace petroleum-based fuels, natural gas pipelines could be used to transmit it, but it would require a network of distribution stations with special storage tanks. In addition, special storages are required at the point of use, such as in motor vehicles, and these are likely to be much heavier than petrol tanks. Hydrogen in this sense will need to be compared with alternatives such as electric vehicles with advanced batteries (including, possibly, vanadium redox) charged from renewable energy.

The future market for hydrogen is still difficult to ascertain. For example, load-following solar thermal electric plants with storage (Chapter 6) could in principle supply both current electricity markets and future electrified transport markets, obviating the need for energy storage such as by more expensive hydrogen systems. However, in some applications, such as intercity air transport, the use of hydrogen fuel has distinct advantages. While a solar-hydrogen fuel system has enormous potential, more research and development is required on the efficient production of hydrogen from renewable energy sources and on hydrogen storage.

Some other renewable sources

There are a number of other renewable energy sources that can be noted (for further general information, see Kaneff 1990; and various chapters in Johansson *et al.* 1993). These have either smaller or more distant potential, or are limited by some other factor in the Australian context. They are mentioned briefly below.

The assessment of Australia's *wave-power* potential has just begun with the publication of a report from Victoria (Hoy *et al.* 1991). Overseas research looks promising, and a small demonstration plant has recently been installed at Islay in the Inner Hebrides, UK. Work done at the Bureau of Mineral Resources suggests that Australia has very large resources for *geothermal* power, in the form of hot dry rocks (Wyborn 1992). According to current knowledge, this resource is mostly located quite far from major population centres, such as in the Cooper Basin in the north-west of South Australia. Being geologically inactive, Australia does not have other geothermal resources such as the steam vents or hot springs that occur in more volcanically active areas.

The potential for large scale *tidal power* is even more distant geographically, being restricted to remote areas, notably Australia's north-west where there is only a small demand for electricity. The utilisation of this potential would require either the creation of huge local industries, or the construction of a very long transmission line to eastern Australia (where there will be vast overcapacity in power stations for the rest of the 1990s) or the local conversion of tidal electricity into hydrogen for export. Large-scale tidal power could also have major environmental impacts on estuaries. *Ocean thermal electric conversion* (OTEC) utilises the energy potential of temperature gradients in oceans. This is potentially a very large energy source, but is as yet quite distant. Finally, the storage and transport of energy, most particularly solar energy, by *chemical* means has been investigated for some time. A typical example is where, for example, solar energy is used to cause a reaction which converts and stores thermal energy as chemical energy, with this being transportable, and then can be used when the reaction is reversed and thermal energy again liberated.

These technologies vary with respect to their current technological and economic status. All are potentially very large contributors to human energy use, but are either some way from being viable or are sources that are geographically distant from concentrations of energy users.

Conclusion

Up to this point in our discussion of sustainable energy systems, a summary has been presented of the actual and potential contribution of renewable energy to future, sustainable energy systems. This is a complex and at times highly technical subject, but the preceding summary should have sufficed to indicate that the potential is both real and significant. Renewable energy offers a great diversity of technologies suited to a wide range of applications, and its environmental impacts are typically far less than those of current fossil fuel and nuclear options. While many renewable technologies require further research and development to become viable, there are options available and viable in the near term that allow a significant program of reform to be initiated. Such support for developing and implementing renewable energy technologies can contribute not only to making Australian energy systems more sustainable, but also to building a potentially lucrative export manufacturing industry. As has been stressed, in many cases the major barriers to adoption of renewable energy are not technical. They are political and institutional, or arise from distorted markets and the inappropriate economic signals that they produce. The pathway forward for renewable energy is threefold: adopting and improving already viable technologies; removing non-technical barriers; and strongly supporting research into and development of the longer term prospects. The challenge is clearly mostly in the political sphere.

References

Ballinger, J.A., Prased, D.K. and Rudder, D. 1992. *Energy efficient Australian houses.* 2nd ed. Canberra: Australian Government Publishing Service.

Blakers, A., Crawford, T., Diesendorf, M., Hill, G. and Outhred, O. 1991a. *Opportunities for the Australian wind energy industry in reducing greenhouse gas emissions.* Report prepared for the Department of the Arts, Sport, the Environment, Tourism & Territories, Canberra. Sydney: UNISEARCH.

Blakers, A., Green, M., Leo, T., Outhred, H. and Robins, B. 1991b. *The role of photovoltaics in reducing greenhouse gas emissions.* Report to the Department of the Arts, Sport, the Environment, Tourism and Territories, Canberra. Sydney: UNISEARCH.

Bockris, J.O., Vizirolu, T.N. and Smith, D. 1991. *Solar hydrogen energy: the power to save the world.* London: Optima.

Danish Ministry of Energy. 1990. *Energy 2000: a plan of action for sustainable development.* Copenhagen: Ministry of Energy.

Deni Greene Consultancy Services. 1991. *Least cost greenhouse planning for Victoria: Part 1.* Report prepared for the Department of Planning, Policy and Landscape, Royal Melbourne Institute of Technology. Sydney: Deni Greene Consultancy Services.

Diesendorf, M. 1993. Power politics in Western Australia. (Letter) *Windpower Monthly.* 9(2): 4.

Ecologically Sustainable Development (ESD) Working Groups. 1991. *Final report: energy production. Final report: energy use.* Canberra: Australian Government Publishing Service.

Edwards, B.P. (circa) 1990. *Opportunities for improved resource use in the Australian sugar industry.* Mackay, Queensland: Sugar Research Institute.

Gray, E.M. 1991. The hydrogen economy. Paper to the ESD Workshop on *Energy Efficiency and Renewable Energy*, Canberra, 16–18 April 1991.

Hagen, D.L. and Kaneff, S. 1991. *Application of solar thermal technologies in reducing greenhouse gas emissions.* Report to the Department of the Arts, Sport, the Environment, Tourism and Territories. Canberra: ANUTECH.

Hoy, R.D., Smith, B.K., van der Riet, P., McCowan, A. and Tyshing, R. 1991. Wave energy potential of the Victorian coast. In: *Proceedings, Solar 91: Energy for a sustainable world*, Vol. 2. pp. 563–570. Caulfield West, Victoria: Australian and New Zealand Solar Energy Society.

Johansson, T.B., Kelly, H., Reddy, A.K.N. and Williams, R.H. 1993. *Renewable energy: sources for fuels and electricity.* Washington DC: Island Press.

Kaneff, S. 1990. Renewable energy. In: Dovers, S. (eds), *Energy options for sustainability.* pp. 48–100. Canberra: Centre for Resource and Environmental Studies, Australian National University.

Lemon, J.H. 1991. Water turbines and hydro-electric plants. Paper to the ESD Workshop on *Energy Efficiency and Renewable Energy*, Canberra, 16–18 April 1991.

Møller, T. 1985. Windpower support is good business for the state. *Windpower Monthly.* 1(10): 4–8.

Reeves, R.R. and Lom, E.J. 1992. *New technology for production and use of ethanol as a transport fuel and lignin as a coal replacement.* Sydney: APACE Research Ltd.

Schodde, R. 1990. *Commissioned study into the prospects for using energy storage in Australia.* Unpublished report to the Department of Primary Industries and Energy on behalf of McLennan Magasanik Associates P/L.

Skyllas-Kazacos, M. 1991. The vanadium redox battery for efficient energy storage. Paper to the ESD Workshop on *Energy Efficiency and Renewable Energy*, Canberra, 16–18 April 1991.

Todd, J.J. and Singline, R. 1989. *The impact of woodheaters on air quality in Australia.* Hobart: Centre for Environmental Studies, University of Tasmania.

Versluis, P. 1991. The market for solar hot water systems in Australia and barriers to adoption. Paper to the ESD Workshop on *Energy Efficiency and Renewable Energy*, Canberra, 16–18 April 1991.

Williams, E.G. 1991. Small hydro power in Australia as an ecologically sustainable development. Paper to the ESD Workshop on *Energy Efficiency and Renewable Energy*, Canberra, 16–18 April 1991.

Wyborn, D. 1992. *Geothermal energy from hot dry rocks in Australia.* Canberra: Bureau of Mineral Resources.

PART FOUR

Towards sustainable energy systems

CHAPTER 8

Economic policies for sustainable energy use

MICHAEL COMMON

As made clear in Chapter 2, the use of extrasomatic energy is a central and pervasive feature of modern economic activity. Policies intended to affect energy use will have repercussions throughout the economy. Equally, policies directed at other sectors of the economy must be presumed to have implications for energy use and the energy sector. Sustainable energy use cannot be divorced from the general problem of sustainability. Much can be learned by looking in detail at particular energy supply and use technologies, the so-called 'bottom-up' approach. But, it is also necessary to start at the other end, the so-called 'top-down' approach, and look at economic activity as a whole in relation to sustainability and energy use.

One way to do this is to consider the case for a particular policy. This chapter considers the case for the carbon taxation of fossil fuels as a policy instrument to promote sustainability. It does not argue that carbon taxation alone is going to solve the sustainability problem: there are roles for other policy instruments. It is argued that carbon taxation should be high on the agenda in any serious debate about policies to promote sustainability. The chapter is organised as follows. First, the nature of the sustainability problem is reviewed. Second, the case for an economy using fossil fuel carbon as a tax base is discussed. Then multilateral agreement on carbon taxation is considered in relation to the enhanced greenhouse effect problem. Finally, some alternative instruments to taxation for pollution control are reviewed, and their possible use for reducing fossil fuel carbon dioxide emissions discussed.

171

Sustainability and energy

Sustainable development is development which meets the needs of the present without compromising the ability of future generations to meet their own needs.   *(World Commission on Environment and Development 1987: 43)*

This is the basic definition of sustainable development given in the 'Brundtland Report', which was responsible for the recent popularisation of the idea. It has been widely noted that this definition is somewhat imprecise, and there is substantial disagreement over what an operational definition would be. The term 'sustainability' also features in the debates arising, and there are questions about its definition, and how it and sustainable development are related. However, it is unnecessary to get bogged down in definitional matters here. The basic nature of the problem which both terms refer to is reasonably clear. It concerns threats to intergenerational equity arising out of the interactions between economic activity and the natural environment. Thus, for Brundtland the problem for which sustainable development is the answer is as follows. There are in the world today many people suffering material deprivation. Alleviation of this poverty is taken to require global economic growth/development. But, many are convinced that further growth/development of the type experienced in the last two hundred years would entail effects on the environmental base for economic activity such that its ability to support such activity in the future would be seriously eroded (see, for example, Meadows *et al.* 1992). On this view, improvements in the lot of the current generation would impose severe costs on future generations.

The perception of economy–environment linkages underlying this view is represented schematically in Figure 8.1. In relation to the economic activities of production by 'firms' and consumption by 'households', the natural environment is shown as serving three functions – waste assimilator, resource base, and source of amenity services. In Figure 8.1, the three boxes representing these functions intersect to highlight an important feature of the situation, which is that the three functions interact. Waste flow in excess of assimilative capacity means pollution, which impacts adversely on the environment's capacity to provide resource inputs to production and amenity services to households. A simple example of what is involved is as follows. Consider a factory discharging its wastes into a river estuary which is also used by a commercial fishing industry and for recreational swimming. If the waste flow exceeds assimilative capacity, the river becomes polluted reducing fish catches and making swimming dangerous to health.

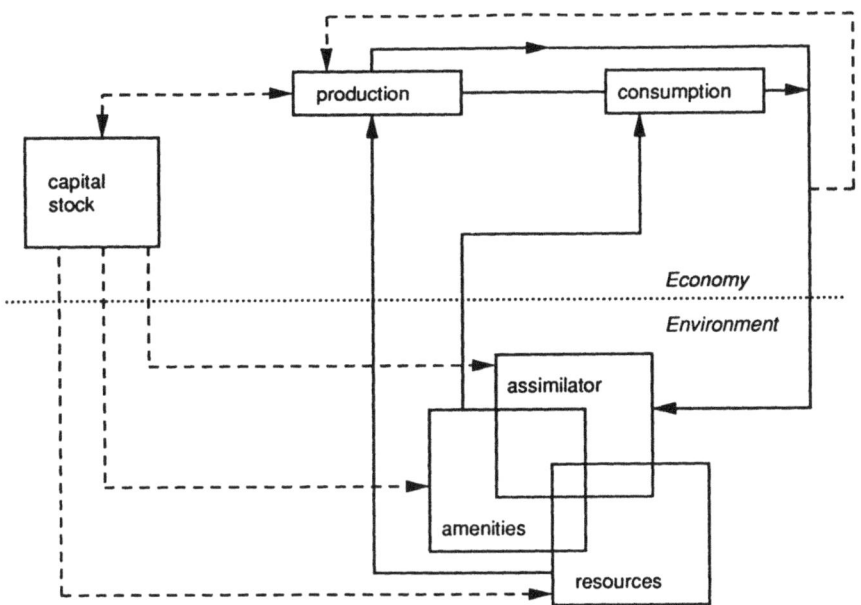

Figure 8.1 Model of economy–environment linkages.

It is conventional to divide resources into renewable and non-renewable categories. Renewable resources are biological stocks that reproduce over time and for which, leaving aside problems of pollution and habitat destruction, there are positive harvest rates that are indefinitely sustainable. The range of sustainable harvest rates depends on the reproductive behaviour of the species involved. Non-renewable resources are physical stocks for which there are no indefinitely sustainable positive use rates. These conventionally distinguished classes of resources are both stock resources: variations in current use rates affect the amounts available for future use. There also exist flow resources, where the current level of use has no implications for future availability: solar radiation is an example of this class of resource.

Current economic activity has the potential to reduce the capacity of the environment to provide useful inputs to, and services for, future economic activity. What Figure 8.1 is intended to make clear is that this is not solely a matter of direct resource depletion. Current use of material resources necessarily means that, via the materials balance principle (mass conservation), wastes discharge into the environment. Where this is at rates in excess of assimilative capacity, there is additional erosion of the resource, and amenity service, base available to the future. Whether this potential intergenerational tradeoff is an actual problem depends, of course, on the levels of current extractions from and insertions into the natural environment. Those, such as Brundtland, who take the view that the problem is now actual are making an empirical as well as an analytical claim. The claim is, essentially, that global extraction and insertion levels are now such that increasing them, to address current poverty and accommodate an increasing population, could in fact result in erosion of the environmental base for future economic activity.

Brundtland does not claim that such erosion is inevitable. On the contrary, the claim is that given certain changes economic activity levels can be increased, and poverty eliminated notwithstanding a growing population, without significantly eroding the environmental asset base. That is, sustainable development is possible. The majority of economists also would deny that further economic growth will necessarily degrade the natural environment's capacity to support economic activity. The essence of their argument is that the presumption that it will overlooks three important substitution possibilities:

  (i) The composition of the production and consumption aggregates can change as the aggregate levels rise, with less environmentally damaging activities being substituted for more environmentally damaging ativities, so reducing the impact of any given aggregate level.

 (ii) Recycled wastes may be substitutable for newly extracted resources, reducing both demands upon assimilative capacities and extraction requirements per unit of production.

(iii) Part of the output from production can be accumulated as capital, rather than consumed, the services of which may be substitutable for resources and environmental services.

Possibilities (ii) and (iii) are indicated by the dashed lines in Figure 8.1. Two issues now arise. First, what is the empirical status of these substitution possibilities? How easily, for example, can accumulated capital replace exhaustible natural resources in production? Second, the existence of such substitution possibilities does not of itself guarantee that they will be exploited.

In regard to the first issue, answers must be sought on a case by case basis. Presumably, some environmental resources and services are more critical, in the sense of being less possible to substitute for, than others. On the whole, economists take the view, explicitly or implicitly, that there is nothing that cannot be substituted for by one or other of the above routes. This view sharply differentiates most economists from many others concerned with the sustainability problem. The difference of views cannot be resolved scientifically: it is not solely a technological question. Most economists would regard virtual reality devices as capable of providing substitutes for the actual experience of wilderness recreation. Clearly, for many people this would not be an acceptable substitution. On the other hand, presumably most people would be indifferent as to whether their motor car was made of newly extracted or recycled materials, assuming no difference in price and performance.

In regard to the second issue, economists look to the price mechanism to induce appropriate substitutions. This is not the same thing as leaving everything to market forces. On the contrary, most of the standard economic analysis of economy–environment linkages is about the ways in which 'market failure' can best be corrected (for introductory expositions, see Common 1988 or Pearce and Turner 1990). Ideally, the appropriate substitutions will be induced by market-based responses to changing relative prices reflecting true relative scarcities. In fact, for many natural resources and environmental services markets do not exist at all, due to the non-existence of relevant private property rights. Where markets do exist, they frequently fail to work as required on account of monopoly power or imperfect information available to market agents. There is, in standard economics, a presumption in favour of correcting market failure by the use of policy instruments which work through relative prices and the incentives that thus arise, rather than through alternative ways of influencing behaviour such as regulatory control. The basis for this presumption is discussed in the final section of this chapter.

Before looking at energy in the context of the schema of Figure 8.1, one further general point needs to be made. This concerns capital accumulation. As noted above, capital accumulation is an alternative use to consumption of the output from production. It requires saving and investment. Whatever the actual technical possibilities for substituting capital for resources and environmental services are, their realisation is possible only as the result of saving and investment. If all of the output from production is consumed, then no capital accumulation is possible. As the level of saving and investment rises, so the price of capital relative to other inputs to production would be expected to fall. This would create incentives to use more capital and less of other inputs, including inputs based on environmental assets.

The actualisation of capital substitution requires the production of particular items of equipment, capital goods of various kinds, and the production of these will itself require consumption of resources and environmental services. Capital goods vary in their requirements for these inputs. If capital substitution is to promote sustainability, it is necessary to encourage not only investment generally but also to encourage investment to go into forms that substitute for scarce resources and environmental services and which themselves are not resource- and environmental service-intensive. To the extent that scarce resources and environmental services command high prices, possibly after policy intervention, incentives will exist to encourage the required composition of investment. Of course, it may be that there are limited opportunities to produce capital goods that are not themselves resource and environmental service-intensive, and which can substitute for scarce resources and environmental services. To the extent that this is the case, capital substitution is not going to be effective in promoting sustainability. Reliable empirical information on such matters is itself scarce. However, it is clear that while a high level of saving and investment does not guarantee that sustainability can be promoted through capital substitution, low levels are a problem for sustainability. The 'Hartwick Rule' says that it is necessary, but not sufficient, for sustainability – defined as a constant level of consumption – that the rent (the difference between sales revenue and extraction costs) arising in competitive natural resource extraction is saved and invested (Hartwick 1977).

Now consider briefly the characteristics of the alternative sources of extrasomatic energy in relation to the schema of Figure 8.1, and the substitution possibilities discussed here. The first point to be made is that energy cannot be recycled, whatever its source. Fossil fuel reserves are non-renewable resources and as such are actually rather than potentially exhaustible. A prospect which is widely discussed, in earlier chapters of this book for example, is the substitution of renewable for non-renewable sources of energy. There are two main reasons for supposing this to be a 'good thing'. First, it addresses threats to sustainability arising from the depletion of the non-renewable sources. Second, it is believed that it would reduce the wastes generated in making available any given level of energy service. While it is true that fossil fuel combustion is a major source of atmospheric pollution, it is not true that all renewables are totally free from problems in regard to waste generation and impact on environmental amenity services. Wood combustion, for example, does release wastes, including carbon dioxide, into the atmosphere. While the operation of solar energy devices involves no stock resource depletion or waste discharges, their manufacture involves both. Environmental amenity problems can also arise: windmill farms are visually intrusive, and hydro-electricity proposals have frequently generated controversy on account of their impact on wilderness areas.

The point here is not that renewable for fossil fuel substitution can never promote the cause of sustainability. It is that there is no such thing as a 'free lunch' (which is the first law of economics and more or less the second law of thermodynamics). Some lunches may, of course, be cheaper than others, and from a sustainability perspective a general presumption in favour of renewable, as opposed to non-renewable, energy sources is very likely justified. However, exceptions cannot be ruled out. The case of nuclear fission is instructive. While uranium is a non-renewable resource, early appraisals saw reserves as lasting an extremely long time, given the possibility of breeder reactors, and providing the prospect of a very low-cost and environmentally benign source of electricity. Experience with the widescale implementation of the technology has provided a very different perspective for the majority of those concerned to promote sustainability.

Energy conservation is largely about the substitution of capital and/ or some other material resource for fuel, as inputs to production and consumption. It is, of course, also possible to substitute human energy input for fuel input, but this is not what most people who advocate energy conservation appear generally to have in mind. Generally, what energy conservation is taken to mean is the achievement of the same level of delivery of some good or service for a smaller input of fuel. A good example is insulating a house to reduce fuel use. Another would be the installation of improved control systems for heating systems. In each case, the essential point is that an investment is undertaken which involves the substitution of other inputs for energy. Again, it needs to be noted that it cannot be generally assumed that the other inputs come free from a sustainability perspective. The manufacture of insulation materials, for example, involves both resource depletion and waste generation.

The case for carbon taxation

The standard way to think about taxation *per se* in economics is to ask the following question. Given that the government must raise some revenue, and given the feasible tax bases and systems, what is the least bad way to do it? Economics has also devoted a lot of attention to the use of taxation as an instrument for environmental policy (see below). There does not appear to have been any work done which takes sustainability as the criterion for choosing between tax systems. In terms of the nature of the sustainability problem as sketched in the previous section, the following would appear to be desirable characteristics for a tax system:

  (i)   It would protect critical environmental resources and processes.
 (ii)   It would encourage saving and investment.
(iii)   It would channel investment into projects providing substitutes for environmental resources and processes.

(iv) It would discourage population growth.
(v) It would promote intragenerational equity.

This is, of course, in the nature of a list of motherhood statements. The sustainability problem is characterised, for example, by the fact noted above that we do not know with any precision which are the 'critical environmental resources and processes'. However, it is necessary to start somewhere. It should be noted explicitly that what is at issue is not a tax panacea for the sustainability problem. It is not a problem that can be addressed by a single instrument. What is at issue here is the question of whether a concern to promote sustainability has any implications for a tax base/system choice, given the need to raise revenue. The following considers the case for some switch from income to carbon dioxide emissions from fossil fuel combustion as tax base, in terms of the sustainability-promoting desiderata listed above.

Environmental protection

A major perceived threat to sustainability is the enhanced greenhouse effect. This exemplifies the uncertainties which attend economy–environment interactions, and the lack of knowledge concerning substitution possibilities as between capital and resource/environmental services. There is now a considerable literature on carbon taxation in relation to the enhanced greenhouse problem: for references, see Pearce (1991) or Industry Commission (1991), for examples.

Fossil fuel combustion's contribution to environmental damage is not confined to atmospheric releases of carbon dioxide. Other atmospheric wastes arising are: particulates, sulfur dioxide, nitrogen oxides, hydrocarbons, and carbon monoxide. Fossil fuel combustion is generally regarded as the major source of atmospheric pollution, and the costs arising have been shown to be substantial: see, for example, Fisher and Smith (1982).

It should also be noted that the environmental impact of fossil fuel combustion is not limited to the emissions from that process itself. Waste discharges arise from the moving and transforming of matter found in the natural environment. The ability to move and transform matter in the absence of extrasomatic energy sources is limited, and waste generation in total has grown broadly in line with fossil fuel use (see, for example, Boyden *et al.* 1990). Higher energy prices are generally understood to increase the incentive to many kinds of recycling.

Fossil fuels are non-renewable resources, and recycling is impossible. Higher fossil fuel prices consequent upon taxation of their carbon content would slow down depletion rates.

On the criterion of creating incentives to protect the environmental asset base, taxing fossil fuels according to carbon content looks good. It directly addresses a perceived major threat to sustainability, the enhanced greenhouse effect, as well as other environmental damages, and resource depletion. But nothing is all good. Carbon taxation would encourage the substitution for fossil fuels of other energy sources with environmental impacts, such as nuclear fission. It may be that the widespread use of nuclear fission would pose a greater environmental threat to sustainability than global climate change. This point is made here to illustrate the uncertainties that attend the sustainability debate, and the consequent need for judgement as opposed to definitive technical resolution. I think it is a reasonable judgement that, on balance, some switch from income to carbon as a tax base would reduce threats to sustainability. Also, as noted above, carbon taxation is not the only available instrument, and its use does not preclude future developments, such as the taxation of energy use from other sources.

Saving and investment

It is widely believed that some switch from income to consumption as tax base would encourage saving and investment, at the expense of current consumption. In Australia, for example, this belief was part of the case for the Liberal/National Coalition's tax reform package (Hewson and Fisher 1991), the major element of which was reductions in income taxation financed by increased consumption taxation. A switch from income to carbon as tax base would be very similar to a switch from income to consumption in this respect. This is because a tax on fossil fuel production or consumption would be passed forward through to the prices of all commodities. The production of all commodities in a modern economy involves the use of energy directly and indirectly: see Chapter 2 for relevant Australian data. A carbon tax would work like a consumption (or goods and services) tax, but with differential rates across commodities according to their carbon intensities and the extent to which the tax could be passed forward.

Project selection

As noted above, from a sustainability perspective, the composition of aggregate investment is as important as the level of the aggregate. Some switch from income to carbon as tax base would alter the structure of input and output prices, thus impacting on project appraisals and the consequent selection of varieties of capital goods. Arguments already

advanced here would suggest a general tendency to favour less environmentally damaging projects, and projects making smaller demands on
the subset of non-renewable resources which are the fossil fuels. In
regard to the former, what can be claimed is a tendency not a guarantee. A particular carbon emission-reducing project could well involve
increased damage to some other environmental services, and higher
depletion rates for some other resource stocks.

Population growth

It seems at first sight that some tax base substitution as between fossil
fuel carbon and income would be neutral in regard to family size
incentives. The relative price effects of carbon taxation do not appear
to have any major implications in this regard, but this requires more
detailed analysis. In some countries the income tax system incorporates
child allowances which reduce taxable income. However, this is not a
necessary feature of income taxation. Its effect could be replicated with
carbon taxation by welfare payments related to family size. The family
size incentive effects of some tax base switching would depend on the
pre-existing situation in regard to income tax details, the size of the
switch, and any accompanying changes to the welfare payments system.
The main point is that a package designed to discourage population
growth would likely be seen by many as undesirable on intragenerational equity grounds.

Intragenerational equity

It is widely believed that increasing prices for fossil fuels would be
regressive in impact, given that the poor spend larger proportions of
their total expenditure on fuel than do the rich. This would imply that
a switch from income to carbon as tax base would be regressive in
impact. Some evidence on this is considered below.

So, on three of these criteria there is a presumption that carbon
taxation, replacing some income taxation, would promote the cause of
sustainability. On the fourth, it appears that it would be appropriate to
presume neutrality, and, on the fifth, that such a switch would generally
be considered undesirable. Some of the issues arising have been investigated quantitatively using economic models, and some of the results
arising are now briefly, and selectively, reviewed. These results did not
arise in models designed to investigate the sustainability problem, but
in conventional economic models used to investigate questions of
environmental protection and government revenue raising.

In its report on the costs and benefits for Australia of action to reduce greenhouse gas emissions, the Industry Commission (1991) used two applied general equilibrium models. 'WEDGE' was used to examine the implications for Australia of varying levels of international cooperative action, and 'ORANI' to examine unilateral action by Australia. Some results arising are given in Table 8.1. In both sets of simulations the models were used to calculate the rate of carbon dioxide taxation that would produce a specified cut in total emissions, with the carbon tax revenue arising exactly offset by reductions in income taxation. The switch in the tax base analysed was, that is, revenue neutral. The Industry Commission interpreted the output losses shown as the gross costs associated with the tax base switch. No attempt was made to estimate any enhanced greenhouse effect reduction benefits arising, so that net costs could not be reported.

Consider first case 4 where Australia is the only country to act to cut carbon dioxide emissions. In this case Australia incurs domestic costs and lost competitiveness in international trade. Also, since Australia's contribution to global emissions is approximately 1.5%, it would be expected that unilateral action on its part would have a negligible effect on the future path of atmospheric concentrations of carbon dioxide. Maximum impact in this regard would arise where all nations act on emissions, case 1 in Table 8.1. In this case, the adverse effect on Australian competitiveness would be ameliorated. However, to the extent that other nations also incurred output losses, this would work to reduce their demands for Australian exports. Case 2 involves only the developed nations which are members of the OECD acting on carbon dioxide emissions. As compared with case 1, the costs to Australia are reduced by 0.03% of NDP, while the impact on concentrations would also be reduced. In case 3, the rest of the OECD acts, but Australia does

Table 8.1 *Industry Commission carbon tax results for Australia*

Case	Carbon cut	Output loss (%)	Tax rate
1	40% Global cut[A]	1.5	34[B]
2	40% OECD cut[A]	1.47	na
3	40% OECD less Australia cut[A]	0.06	na
4	44% Unilateral Australia cut[C]	2.1	21.75[D]

Source: Industry Commission 1991.
Notes: [A] WEDGE result for Real Net Domestic Product, database 1988.
     [B] US$ (1988) per tonne carbon dioxide.
     [C] ORANI result for Real Gross Domestic Product, database 1980–1.
     [D] A$ (1988) per tonne carbon dioxide: revenue $4.7 billion.

not. This case illustrates the incentives to 'free-ride' in relation to the enhanced greenhouse effect problem. As compared with case 2, there is a large reduction in the costs borne by Australia, but Australia would still enjoy any benefits on account of reduced global warming consequent on OECD action. Such benefits would be very little reduced on account of Australia's inaction.

The values reported in Table 8.1 come from static, rather than dynamic, models. This means that the models cannot pick up any aggregate saving and investment effects arising from the tax base movement modelled. Also, no credit can be given for any non-greenhouse environmental benefits arising from the reduced fossil fuel use, such as reductions in other pollutants. The models do allow for capital-energy substitution, labour-energy substitution, and inter-fuel substitution (somewhat crudely), but the technology is fixed, so that changes in relative prices cannot influence the direction of technological change. Also, the models assume that all production activities are conducted in cost minimising ways, given the technologies modelled: many commentators argue that this is an inappropriate assumption in regard to energy use (see Appendix F, Industry Commission, 1991). On these grounds one may say that the model results are likely to overstate the costs of meeting the carbon dioxide emission reductions targets. On the other hand the models do not capture adjustment cost effects, do assume a highly flexible economy, and do not capture any effects on intragenerational equity within nations.

Carbon taxes would raise fuel prices, and there is a widespread view that this would have adverse implications in regard to intragenerational equity. Table 8.2 reports some results on this for Australia. Input–output and energy data for 1986–7 were used to calculate the impact of carbon taxation at \$20 per tonne of carbon dioxide (approximately the Industry Commission rate for a unilateral 44 per cent emissions cut, see Table 8.1) on the prices facing households, and Household Expenditure Survey (HES) data for 1984 to map this into consumer price index (CPI) increases for households by decile (1984 HES data were used because 1984 was the most recent year for which expenditure patterns were available by decile). In Table 8.2 the second column shows CPI results when the effect of carbon taxation on the prices of all (27) commodities are accounted for, while the third column shows CPI results when only the effects on the three fuel commodities are accounted for. The results show that carbon taxation in Australia would be regressive, but less so than much commentary would suggest. They also show that regressivity is overstated if only the effects of carbon taxation on the fuel commodities are considered. As discussed in Chapter 2, the production of all commodities uses fossil fuels indirectly as well as directly, and the results in the second column of Table 8.2 pick up the implications of this for the prices of all commodities.

The results in Table 8.2 do not allow for any substitution responses in production or consumption, nor do they allow for any effects arising from the use to which the carbon tax revenue is put. In regard to the first point, we would presumably want to assume that substitution responses in consumption could only reduce the impact on any household group, with commodities for which the tax-induced price increases were smaller being substituted for those where they were greater. This could conceivably increase regressivity as measured in Table 8.2, if better-off households did more substitution than the less well-off. The distributional implications of allowing substitution responses in production are more complex. Some relevant applied general equilibrium results are noted shortly. One possible implication can be noted here. The empirical literature generally reports labour and energy as substitutes in production. This would imply a presumption that, by favouring the use of labour over energy, carbon taxation would increase employment for a given level of output. This is significant in that unemployment is a major source of social inequity.

Two applied general equilibrium model studies for the USA have allowed for substitution responses in consumption and production, and distinguish groups of households by income level. Boyd and Uri (1991) analyse the imposition of a tax of 10 cents per BTU $\times 10^6$ (a little over 1 GJ) across all energy. They find that household welfare declines by an amount varying from 'almost one and one-half to two times the amount

Table 8.2 *CPI increases by decile for tax of $20 per tonne CO_2*

Feature		Accounting for all commodity price increases (%)	Three fuel commodity price increases only(%)
Decile	1	2.885	1.534
	2	2.995 (high)	1.657 (high)
	3	2.974	1.604
	4	2.850	1.444
	5	2.876	1.452
	6	2.774	1.353
	7	2.804	1.313
	8	2.774	1.278
	9	2.666	1.164
	10	2.621 (low)	1.097 (low)
All households		2.785	1.311
Ratio high/low		1.14	1.51

Source: Common and Salma 1992.

of the gain in revenue' (p. 271). The size of household loss is not correlated with income level, so that the energy tax is not regressive in impact. Boyd and Uri's model is static, and environmental benefits are unaccounted for. Boyd and Krutilla (1991) compare the costs of raising a target amount of revenue by income taxation, and by coal and oil taxation where the rates on the two fuels are differentiated (in the ratio 1.6/1) to reflect 'the relatively high environmental cost associated with coal production and consumption' (p. 9). It is assumed that the additional revenue is used to increase government spending. For additional revenue of $100 billion, they find that the impact of the energy tax is to reduce carbon dioxide emissions by 52 per cent, for a welfare cost some two and a half times that of the income tax. Distributionally, the energy tax impact is found to be mildly regressive, that is, households lower in the income distribution suffer a little more than those higher up the income distribution. Again, the model is static and no environmental benefits are accounted for.

It should be emphasised that the results reported here come from models which are simple approximations to a complex reality, and that the parameter values used in the models are in the nature of 'guesstimates'. In its report, the Industry Commission was very explicit about this, stating that:

> The Commission is nevertheless conscious of the inevitable limitations of any attempt to model such complexity. The task stretches current modelling capabilities to the limits ... the results reported should be regarded as more illustrative than definitive ... Such numbers can only have 'ball park' significance at best and rest on many simplifying assumptions spelled out in the report.
> *(Industry Commission 1991: 3–4)*

Ideally, the reporting of results from applied general equilibrium modelling studies of this type would involve extensive sensitivity analysis over a wide range of variations in parameter values. However, the models themselves are quite complex and involve large numbers of parameters, so that this is difficult to do. Common and Salma (1992) and Barrett (1991) do report the results of sensitivity analyses in much simpler models, where only substitution possibilities in production are allowed for in a highly aggregated way, by using price and income elasticities of demand for fuels. These results show that the tax rates calculated as necessary to achieve given carbon dioxide emissions reductions, and hence the costs of meeting such targets, are very sensitive to the values assumed for the parameters. One would expect this sensitivity to carry over to the more complex models of the applied general equilibrium type as used by the Industry Commission (1991), Boyd and Uri (1990) and Boyd and Krutilla (1991) discussed above.

Global carbon taxation

Sustainability is essentially a global problem in all of its dimensions. The enhanced greenhouse effect problem is a particular threat to global sustainability, and itself has global dimensions in so far as all nations would be affected by any climate change, albeit differentially, and effective preventive action would require concerted action by all nations. Such action would impact differentially across nations in terms of its effects on economies. Recognition of this is an important barrier to securing international agreement on concerted action. The issues arising can be explored using an applied general equilibrium model of the type discussed above which looks at the world economy rather than that of a single nation. The results obtained from such world models are subject to the caveats discussed above for national models.

Whalley and Wigle (1990) used an applied general equilibrium model of the world economy to examine the implications of alternative carbon tax regimes to secure a 50 per cent reduction in global carbon dioxide emissions arising in fossil fuel combustion. Table 8.3 reports some of their results in terms of the costs measured as percentage GDP changes. The world is modelled as comprising six trading economies, and three different carbon tax regimes are examined. Under Option 1, each economy imposes that rate of taxation on the carbon content of its fossil fuel production that is required to cut its emissions of carbon dioxide by 50%. Option 2 also involves each economy cutting emissions by 50%, but in this case the tax base is the carbon content of fossil fuel consumption rather than production. Under Option 1, tax is levied on fossil fuel production, while, under Option 2, it is levied on fossil fuel consumption. When each economy is acting to cut its own emissions by the same target amount, this is an important distinction. Option 3 is where there exists some global tax collecting agency, which sets the carbon tax rate at that required to reduce global emissions by 50%. In this case, each economy does not necessarily cut emissions by 50%. The cut in an individual economy depends on the costs to it of cutting emissions in comparison with the costs to it of paying tax revenue to the international agency. In this case, the outcome is the same irrespective of whether fossil fuel production or consumption is adopted as the tax base. The international agency returns the revenue that it collects to nations in proportion to their population size, that is, per capita disbursements are equalised.

The important features of these results are as follows. First, world total costs are relatively insensitive to the choice of tax option, but are lowest for Option 3. This is consistent with one of the major propositions of the economic theory of instrument choice for environmental

Table 8.3 *Global carbon taxation: percentage GDP changes*

Region	Tax option		
	1	2	3
European Community	−4.0	−1.0	−3.8
North America	−4.3	−3.6	−9.8
Japan	−3.7	−0.5	−0.9
Other OECD	−2.3	−2.1	−4.4
Oil exporters	+4.5	−18.7	−13.0
Developing/centrally planned	−7.1	−6.8	+1.8
World	−4.4	−4.4	−4.2

Source: Whalley and Wigle 1990.
Note: Results are for a global 50 per cent reduction achieved by:
   Option 1 is national fossil fuel production taxation;
   Option 2 is national fossil fuel consumption taxation;
   Option 3 is uniform global taxation.

protection, to be discussed below. Basically, the point is that if all economies face the same rate of tax, but have different costs associated with emissions reduction, each economy will reduce by different amounts, with those where it is cheapest reducing the most and vice versa. This means that the global reduction will be loaded across economies so that the global total cost of reduction is minimised. The second point to be made concerns the distribution of costs across regions as between the three options. In general, there is no reason to suppose that the loading of emissions reductions that minimises total costs will be equitable as between economies. However, in the case being considered here, it turns out that, given the way the carbon tax revenues are distributed, efficiency and equity go together. Under Option 3, the 'Developing/centrally planned' economy actually gains in GDP terms as a result of the tax collection and disbursement, while all the other, basically rich, economies suffer losses of GDP.

Alternative instruments

In the economics literature which deals with the choice of policy instruments for environmental protection in regard to pollution, four alternative classes of instruments are usually distinguished:

(i) Manipulation of the cultural environment. Particular forms include:
   • publicity to generate social pressure on polluters;
   • education of actual and potential sufferers;

- education on environmental functioning;
- financing research on pollution and the environment;
- facilitation of pressure group activities.
(ii) Regulation, or direct control, of waste discharger behaviour. This can take two main forms:
   - specification of allowable emissions quantities;
   - specification of process and/or equipment.
(iii) Price incentive modification. Four principal options exist here:
   - emissions taxation;
   - payment for emissions reduction, that is, subsidisation;
   - creation of markets in emissions permits;
   - input taxation.
(iv) Public provision of waste treatment facilities.

The first class of instruments is sometimes referred to as 'moral suasion', and the second as 'quantity control' or 'command and control'. The various instruments are not mutually exclusive. Regulation of process/equipment could, for example, be combined with payment in respect of costs arising, and adopted concurrently with a publicity campaign concerning the pollutant in question. Again, emissions taxation could be used to finance public provision of treatment facilities for the taxed emissions. Where the rate of taxation is set for cost recovery, rather than provide incentives to standard attainment, the tax is often referred to as a 'user charge'. In principle, carbon dioxide taxation could take the form of emissions taxation, so that tax liability would be assessed on the basis of monitoring of emissions. In practice it would, as discussed above in relation to the modelling studies, take the form of either the taxation of fossil fuel purchases at rates reflecting carbon contents, consumption taxation in the terminology of Table 8.3, input taxation in the terminology immediately above, or the taxation of fossil fuel sales by producers thereof, production/output taxation.

Now consider the problem of an environmental protection agency (EPA) seeking to reduce waste discharges of some kind from all sources in a given jurisdiction by some amount. Initially consider the problem as relating to a national EPA seeking to control emissions arising within national boundaries: the problem of international control of carbon dioxide emissions will be returned to below. The options open to the EPA are as listed above. In order to discuss choice from among these it is necessary to specify some criteria for instrument performance. While there are many criteria that an EPA, or those legislating its powers, could take into account (see Common, 1990, or Bohm and Russell, 1985), economic analysis focuses mainly on:

(i) *Dependability*. This concerns the degree to which it is sure that use of the instrument will result in the realisation of the target level of emissions reduction.

(ii) *Efficiency.* An efficient instrument is one which does the job at the lowest possible cost.

(iii) *Informational requirements.* How much information must the EPA have in order to be able to use the instrument effectively?

A full analysis of the properties of the alternative instruments against even this restricted criterion set is beyond the scope of this chapter (see Common, 1988, Chapter 5, for an introductory treatment), but the main conclusions emerging are as follows:

(i) Moral suasion places minimal information requirements on the EPA, but is neither dependable nor efficient.

(ii) Quantity controls which specify allowable emissions quantities are dependable but not efficient, and do not require that the EPA has information about the production possibilities of the sources it seeks to control.

(iii) Taxes on emissions are efficient and dependable if the EPA does have full information about such production possibilities. In some circumstances efficiency and dependability also apply, given the same EPA information, to the taxation of either particular inputs or the taxation of output (as in the case of carbon dioxide emissions from fossil fuel use). However, typically an EPA would not have the information required to render the use of taxes of any kind dependable. Instead it would have to rely on modelling studies, of the type discussed above in connection with carbon dioxide control, to compute the tax rate which would result in the target level of reduction being achieved. Since the models are at best approximations of reality, the chances of computing the correct tax rate are slight. The tax rate imposed will result in some reduction in emissions, and, whatever that reduction is, it will be achieved efficiently. The tax rate could be adjusted in the light of experience, but this iterative process will involve additional costs.

(iv) Tradeable permits in emissions will be dependable and efficient even if the EPA does not have information on the production possibilities of emissions sources. The EPA simply creates permits in the quantity that corresponds to the target reduction it seeks. Since emissions are prohibited except where sanctioned by permit ownership, target achievement is guaranteed (assuming compliance with the law). It is the fact that the permits are tradeable amongst sources that gives rise to efficiency. Each source will weigh the cost of acquiring an additional permit against the cost of reducing emissions, and trading will move permits from where they have low value, where emissions reductions are less costly, to where they have high value, where emissions reductions are more costly. As a result, the total emissions reduction will be allocated across sources so as to minimise its total cost.

On this basis, the options would rank: 1, tradeable permits; 2, taxes; 3, quantity controls; 4, moral suasion. In fact, in the history of pollution control to date, price incentive instruments (tradeable permits and taxes) have been relatively little used. There are a number of reasons for this, as discussed, for example, in Common (1990). Also, it should be noted that price incentive systems have become more popular in recent years (see: OECD 1989). One aspect of policy instrument choice not covered in the discussion above is that of equity. This is a complex question in full, but two aspects of the matter are relatively straight-forward. First, taxation generates revenue. This can be used to offset any adverse equity effects arising from the impact of the taxation. Whether full offsetting is achieved depends upon the amount of revenue generated and the way it is used. Second, in the case of trade-able permits much depends on how the permits are initially allocated and for how long they remain valid. Permits may initially be issued free or auctioned. If issued free, the question arising is on what basis? Clearly, free issue of permits valid indefinitely in proportion to existing source emissions levels would have very different revenue and equity implications to the case of permits valid for just, say, five years and auctioned initially.

This brief review provides some background for a discussion of the alternative instruments that have been canvassed for use in an inter-national agreement to reduce carbon dioxide emissions from fossil fuels. These emissions are seen as a major factor in the enhanced greenhouse effect problem, which is in turn seen as a major threat to global sustainability. In 1988, an international conference in Toronto proposed a reduction in global carbon dioxide emissions of 20%, on 1988 levels, to be achieved by 2005. The matter has since then been the subject of a great deal of analysis and political debate. To date, no global target for emissions reduction has been adopted by the inter-national community. The following consideration of the dimensions of the problem of instrument choice in this context throws some light on the difficulties attending the realisation of an international consensus on a global target for carbon dioxide emissions reduction.

Quantity control

This would involve each nation being required to cut back on emissions by a given proportional amount from some base year level, or to emit up to a certain absolute amount. The analysis reviewed above indicates a presumption that choice of this class of instrument would not be efficient. This choice would have the property of dependability – the global target would be met. Many take the view that greenhouse problem characteristics require that this property is accorded high priority.

If national targets were internationally agreed, there remains the question of the means by which a nation would seek to realise its own particular target. This question could itself be decided as part of the international agreement, or left for decision by the individual participating nations. The latter option would involve nations in a smaller perceived sacrifice of national sovereignty than the former. This could be expected to make it the more likely outcome. Under this outcome, individual nations could adopt quantity regulation or price incentive systems, and the discussion above applies to each nation.

The central question in negotiating this type of international consensus would be the determination of the national targets. It seems likely that perceptions of equity as between nations would dominate consideration of this question. The simple approach of equal proportional cutbacks across all nations would penalise nations already fossil fuel efficient and the less developed nations, it has been argued. The results in Table 8.3 above for the first two options correspond to equal proportional cutbacks across nations, where each nation chooses, or is required, to use a tax instrument. An alternative suggestion that has been made is that the initial standard should relate to allowable global emissions which would then be shared among nations on a per capita equality basis. This would be seen as favouring less developed as opposed to developed nations. Obviously, many variants on these simple allocations are conceivable. It has been argued that international agreement on national targets would be impossible to reach (see, for example, Grubb 1990).

Taxation

An international agreement to tax carbon dioxide emissions at a uniform worldwide rate could take two forms. The first would involve the tax being levied by an international agency, as in the case of Option 3 in Table 8.3 above. The second would involve it being levied by nation states. In both cases taxation would be globally efficient – that is, least cost – but not dependable. The common global tax rate would realise some emissions reduction at the least cost, but it would not guarantee the attainment of the global target for emissions reduction.

Taxation by an international agency would mean revenues accruing to it, and if the tax were set at rates intended to realise significant global emissions reductions these revenues would be substantial: see Grubb (1990) for some figures on revenues, and Table 8.3 above for the implications of the distribution of revenues in one particular way. In terms of the prospects for achieving international consensus, this gives rise to problems and possibilities. The problems concern the perceived loss of sovereignty by nation states which would be involved in the creation of an international body with significant spending power outside of their influence. The possibilities concern the related questions of equity and inducement to participate in any international consensus. While taxation is efficient it is not necessarily equitable. Rules could be negotiated according to which the international agency would disburse its revenues which would promote equity. A rule could involve, for example, countries receiving a share of total revenue dependent on population size and per capita national income. This would favour large, less developed nations such as India and China, and might be expected to encourage their participation which would otherwise likely not be seen by them as being in their interest. Of course, such a revenue sharing rule would work against the interests of nations such as the USA, and to that extent might discourage their participation. Table 8.3 above illustrates how distribution on a simple equal per capita basis would have similar implications.

The problems of perceived national sovereignty losses and revenue disbursement could be avoided by a form of consensus which had the common tax rate across nations levied by nation states, which would retain the revenue arising. This would also involve the loss of the possibility of promoting equity and encouraging participation by some key players. An additional consideration arises with this form of international consensus if the advantages of fossil fuel input taxation, as opposed to emissions taxation, are sought. This is that the distributional implications across nations then differ according to whether production or consumption is used as the tax base (see Table 8.3).

Tradeable permits

The analysis for a national EPA reviewed above shows that tradeable emissions permits are both efficient and dependable. This carries over, in principle, into the international context and would seem to make them preferable to either quantity regulations or taxation.

The first point to note is that the carry-over is assured only if permits are freely tradeable globally between individual emissions sources, so that the world context is the exact analogue of the nation state. The carry-over is not assured if permits are tradeable only between nation states, since a nation state could choose to meet the emissions quantity it holds permits for by quantity regulation within its borders. In this case, dependability at the global level holds, but full efficiency does not. It would seem intuitive that permits tradeable between nation states would offer efficiency advantages over international quantity controls whatever form of control nation states adopted within their borders.

However, this does not appear to have been demonstrated analytically. In principle, an international consensus could involve a two-stage commitment to tradeable permits – tradeable national permits to be subdivided within nations by tradeable individual permits.

It would appear that tradeable permits would be seen, in whatever form, to involve less sacrifice of national sovereignty than would internationally administered taxation. The intragenerational equity implications would depend primarily on the initial allocation of permits. It is difficult to see how this could be negotiated by nation states except on the basis that permits attach to nation states rather than to individual emissions sources. As with internationally administered taxation, there are problems and possibilities here. The initial allocation would be contentious, but would offer opportunities for addressing existing inequities as between states and for creating incentives for some states to participate. It has been suggested, see Grubb (1990) for example, that initial national allocations based on equal per capita shares of total allowable emissions would serve the cause of international equity and thereby generate incentives for participation.

Clearly, consideration of the choice of policy instrument raises many complex issues (for further analysis and discussion see Grubb 1990; Ingham and Ulph 1991; or Industry Commission 1991). It is clear that a purely technical analysis cannot provide a uniquely 'correct' answer to the question of which is the best policy instrument. The matter involves judgement as well as analysis. This particular problem is, in its complexity and its multi-faceted character, an exemplar of the sustainability problem of which it is a component. Standard economic methods of policy analysis may be of limited usefulness in such contexts. Rather than seeking to identify the 'best' policy instrument, it may be more appropriate to look for instruments that (i) are politically feasible, (ii) meet some threat to sustainability, and (iii) themselves do not involve major threats to sustainability. On this approach, it appears to me that, in developed economies at least, fossil fuel taxation in relation to carbon content does well against (ii) and (iii), and may do well against (i) when properly tested by informed political debate.

References

Barrett, S. 1991. Global warming: the economics of a carbon tax. In: Pearce, D. (ed.), *Blueprint 2: greening the world economy*, pp. 31–52. London: Earthscan.

Bohm, P. and Russell, C.S. 1985. Comparative analysis of alternative policy instruments. In: Kneese, A.V. and Sweeney, J.L. (eds), *Handbook of natural resource and energy economics*, Vol. I, pp. 395–460. Amsterdam: Elsevier.

Boyd, R. and Krutilla, K. 1991. *Energy taxation as a revenue-raising strategy: a general equilibrium analysis*. Paper presented at European Association of Environmental and Resource Economists Conference, Stockholm, June 1991.

Boyd, R. and Uri, N. D. 1991. The impact of a broad based energy tax on the US economy. *Energy Economics*. 13: 258–273.

Boyden, S., Dovers, S. and Shirlow, M. 1990. *Our biosphere under threat: ecological realities and Australia's opportunities*. Melbourne: Oxford University Press.

Common, M. S. 1988. *Environmental and resource economics: an introduction*. London: Longman.

Common, M. S. 1990. Policy instrument choice. In: Common, M. S. and Dovers, S. (eds), *Moving toward global sustainability – policies and implications for Australia*, pp. 87–116. Canberra: Centre for Continuing Education, Australian National University.

Common, M. S. and Salma, U. 1992. *An economic analysis of Australian carbon dioxide emissions and energy use*. End of grant Report to the Energy Research and Development Corporation, Canberra.

Fisher, A. C. and Smith, V. K. 1982. Economic evaluation of energy's environmental costs with special reference to air pollution. *Annual Review of Energy*. 7: 1–35.

Grubb, M. 1990. The greenhouse effect: negotiating targets. *International Affairs*. 66(1): 67–89.

Hartwick, J. M. 1977. Intergenerational equity and the investing of rents from exhaustible resources. *American Economic Review*. 66: 972–974.

Hewson, J. and Fisher, T. 1991. *Fightback! Taxation and expenditure reform for jobs and growth*. Canberra: Liberal and National Parties.

Industry Commission. 1991. *Costs and benefits of reducing greenhouse gas emissions*. Industry Commission Report No 15. Canberra: Australian Government Publishing Service.

Ingham, A. and Ulph, A. 1991. The economics of global warming. In: Bennett, J. and Block, W. (eds), *Economics and the environment: an Australian reconciliation*, pp. 223–248. Perth: Institute of Public Affairs.

Meadows, D. H., Meadows, D. L. and Randers, J. 1992. *Beyond the limits: global collapse or a sustainable future*. London: Earthscan.

Organisation for Economic Cooperation and Development. 1989. *The application of economic instruments for environmental protection in OECD member countries*. Paris: OECD.

Pearce, D.W. 1991. The role of carbon taxes in adjusting to global warming. *Economic Journal*. 101: 938–948.

Pearce, D. W. and Turner, R. K. 1990. *Economics of natural resources and the environment*. London: Harvester-Wheatsheaf.

Whalley, J. and Wigle, R. 1990. *The international incidence of carbon taxes*. Paper at Conference on 'Economic Policy Responses to Global Warming', Instituto San Paolo di Torino, Rome, October 1990.

World Commission on Environment and Development. 1987. *Our common future*. Oxford: Oxford University Press.

CHAPTER 9

Towards sustainable energy systems

IAN LOWE

As energy is a key input to modern society, a sustainable pattern of energy supply and use is absolutely crucial to the development of a sustainable society. This concluding chapter draws on the foregoing analysis to suggest a realistic strategy to develop that desired goal: a pattern of energy supply and use which would be genuinely sustainable, powering a sustainable pattern of social and economic development. It also discusses the political obstacles to implementing the strategy in contemporary Australia.

The broad social role of energy

As discussed in Chapters 1 and 2, modern society is energy intensive. The scale of energy use reflects the vital input of fuel energy to every aspect of modern society. As well as the direct and visible use of fuels, in motor vehicles or in the home, energy is required to supply our shelter, our food, our work, our recreation and so on. An understanding of the broad social role of energy is a crucial background to analysis of the impacts of prospective changes in fuel use.

Energy use is often portrayed as if it were dominated by economic considerations, with occasional concessions to such variables as the climate. In practice, energy demand is very complex; it is significantly influenced by attitudes, values and various aspects of socially dominated behaviour (Stern and Aronson 1984). Large price increases, such as the rises in the sale price of oil fuels in the second half of 1990 at the time of the Gulf War, produce small changes in energy use. For other forms of energy, such as electricity or town gas, consumers often do not have even crude price information at the time of use, so they are clearly not making economic judgements about the marginal utility of competing calls on their finances.

It is very easy to show the effect of lifestyle on energy use by international comparisons (International Energy Agency 1989). This shows, for example, that Australians use about twice as much energy per head as New Zealanders and about half as much as Canadians. Some of the differences between nations can be attributed to climate, but such differences certainly do not account for the bulk of the variations in energy use; there is no such reason for Australian energy use to be double that in New Zealand. Economic factors only account for a small part of the variations in energy use. France has about the same GDP per head as the Netherlands, but only uses about two-thirds as much fuel energy per head. The USA has about the same wealth per head as Switzerland, but uses about three times as much energy per head. The structure of the economy can clearly have an effect; for example, Australian energy use is increased by the government policy of encouraging energy intensive extraction and processing of minerals. Brain-based industries use less energy per unit of economic output.

Chapter 4 shows that Australian transport fuel use is inflated by the strategy of orienting urban transport around the private car, as this uses about twice as much fuel per passenger kilometre as public transport. Car dependence also leads to a pattern of urban development in which people travel longer distances from home to work. So the policy choice of promoting car travel at the expense of public transport has a major impact on energy use. Other public policies clearly affect energy use. For example, the governments in most OECD countries took steps to encourage more efficient use of fuel (especially oil) during the 1970s. Transport policies clearly influence the use of transport fuels. Energy use is also significantly influenced by government policies in such areas as building codes, appliance standards, communications, tourism and so on.

At a wider level, broad economic strategy clearly affects energy use, since the overall pattern of economic development determines the energy used in productive enterprises. An economy oriented toward the extraction and processing of minerals is inevitably energy intensive, whereas an economic strategy of emphasising the provision of services would require much less energy per unit of economic output. The wealth of a society directly affects the level of spending on goods and services, the provision of which in turn requires fuel energy. Thus, energy is used both to create wealth and in the supply of the goods and services purchased with that wealth, so there is a broad association between the level of economic growth and the demand for fuel energy. The link is a complex one because different patterns of wealth production require different levels of fuel use, while different ways of spending the wealth produced will also have different energy needs. In general, wealthier societies use more energy, so the rate of economic growth influences energy demand.

The role of energy in modern industrial society was summarised by Dovers (Chapters 1 and 2). It is a necessary input to all human activity, with the scale of energy use reflecting the scale of physical manipulation of the natural world. The vast bulk of our energy use comes from non-renewable sources. The waste products from using energy are the dominant cause of air pollution. They are also major contributors to changes in the composition of the global atmosphere, leading to concern that human actions may be enhancing the natural 'greenhouse effect' which allows the planet to retain enough heat for life to be possible. These issues all point clearly to the need to consider the sustainability of the current pattern of energy use.

Sustainable development

A good definition of sustainable development is the simple one of the Brundtland Report (World Commission on Environment and Development 1987). It said that sustainable development is a pattern of activity that meets the needs of the present without compromising the ability of future generations to meet their needs. It is salutary to note the Brundtland observation that unless economic decisions are *ecologically* rational we will be unable to maintain living standards, still less improve them. For an activity to be sustainable it must not deplete natural resources significantly, it must not have serious impacts on the natural environment, and it must not prejudice social stability. The first two criteria are obvious; if we deplete resources or seriously alter the natural environment, we clearly prejudice the ability of future generations to meet their needs. The third criterion is important because it is easy to imagine policies that would slow down the rate of resource use or environmental impact, but which would be a clear threat to social stability. For example, Dovers argues (Chapter 2) that much of current energy use is not sustainable; energy use for urban transport is an obvious example. It might be argued by those who have faith in market forces that the way to reduce transport fuel use in cities towards a sustainable level would be to increase the price to such prohibitive levels that many people would be unable to afford to drive long distances. As a general rule, those who live furthest from the centres of our cities travel the longest distances, have the least disposable income, are least likely to have public transport options, and have the least efficient cars. So restricting transport by the price mechanism would have very serious consequences on social stability, particularly in the outer suburbs of sprawling Australian cities. Such a policy would probably not be politically sustainable.

Similarly, it is possible to imagine a policy which would be ecologically sustainable within one region or even one nation, but which could bring that region or nation fundamentally into conflict with its neighbours. For example, if Australia had a long term policy for coping with the run down of oil production from Bass Strait, it might be based on using the oil that will probably be found under the Timor Gap. Indonesia could well be expecting to use that oil. Conflict over oil could easily lead to friction or even armed struggle.

In aiming for sustainable development, we need to take account of the global dimension. As the scale of the human population and the level of use of resources has expanded, environmental issues have been transformed from being local problems to regional issues, then to problems of a global nature.

Resources

The issue of depletion of resources is a complex one. It is influenced, for example, by economics. A price increase, whether driven by scarcity or achieved by control of the market, is likely to make marginal resources economic and so expand the resource base. It is also influenced by technology; a new extraction technique may turn mineral deposits which were previously inaccessible into useful resources. It is influenced by the political culture of the time, as shown by such questions as whether the sand minerals of Fraser Island or the minerals under Coronation Hill should be regarded as resources. These issues are influenced by the political trade-off between the economic benefits of exploiting the deposit and the various costs, both social and environmental. Those complications mean that resources cannot be seen in absolute terms; what is a resource at any given time is the product of a mosaic of factors. Few resources are likely to be scarce in the foreseeable future, though there is one particular problem. For the best part of thirty years now, it has become steadily more difficult to find oil (Lowe 1986). It is taking more and more exploration to find new reserves. We are now getting oil from such unlikely places as the North Sea and Alaska. There is pessimism about the chance of ever again finding the sort of massive deposits which have fuelled the expansion in road and air transport since 1950. The current rate of using petroleum fuels does not just threaten the global atmosphere; it also risks depleting oil fields within the lifetime of today's young people. Production from the USA and the former USSR is in rapid decline, making the world more dependent on oil from the Middle East (Flavin 1992). In general, even when we are not depleting resources in absolute terms, we usually exploit the richest and most easily accessible deposits. Thus our activities condemn future generations to a greater economic or environmental cost for energy minerals, even when there is not an unreasonable level of depletion in absolute terms. Market prices do not reflect costs to future generations, so leaving the exploitation of natural resources to the market will always tend to reduce opportunities for our descendants.

Ecological issues

What are known as environmental issues should probably be called ecological issues. The word 'environment' alludes to surroundings. Natural ecosystems are not just our surroundings; humans are part of the natural systems of the planet, so we need to think about the impacts of our actions on those systems. That wider awareness is developing. The whales of the southern ocean, the rainforest of the Amazon and the Great Panda of China are not parts of our immediate environment; concern for the loss of pandas or of forests represents growing awareness of the ecological impacts of human actions. We now face global problems which require global cooperation of an unprecedented kind.

The problem of leaving this to market forces is demonstrated by the depletion of the ozone layer caused by CFCs. The scientific work showing that CFCs were accumulating in the air was published early in 1973, and by mid-1974 the conclusion that the ozone layer would be depleted by continued release of CFCs had been published in the scientific literature. It took another thirteen years to persuade politicians and bureaucrats to interfere in a profitable industry, even for something as serious as depletion of the ozone layer. The uncertainty of the science was an excuse for inaction. Even in the 1990s, we were still hearing arguments that the withdrawal of ozone-depleting substances should not be so rapid as to cause damage to profitable businesses.

The enhanced greenhouse effect is a more intractable problem. We do not actually need CFCs for any purpose at all; there are substitutes, even if they may cost more or not have as good a level of performance. On the other hand, the main human source of the enhanced greenhouse effect is the release of carbon dioxide from the burning of fuels. Because carbon fuels have been cheap to obtain, we have evolved a lifestyle which depends on massive use of these compounds. Scaling down use will be hard to do. Some changes will be expensive. The improbability of the changes which are needed being driven spontaneously by market forces is so obvious that it needs little elaboration.

It is important to recognise the scale of the problem. Worldwide, about 6000 million tonnes of carbon are added to the atmosphere each year by the burning of fossil fuels. The Australian government has adopted as an interim target the goal suggested by a 1989 conference in Toronto, to reduce the level of emissions of greenhouse gases to 20% below the 1988 level by the year 2005. The *scientific* assessment suggests this to be a minimum target, given that the reductions needed to stabilise the composition of the atmosphere are actually much greater. The Intergovernmental Panel on Climate Change (IPCC 1990) said that it would require a 60% reduction in emissions of carbon dioxide and other long lived greenhouse gases to stabilise their levels in the atmosphere. It is therefore urgent to devise practical ways of reducing our impact on the atmosphere. Australia currently emits more carbon dioxide per head than most other countries. Given that there is a reasonable expectation in Third World countries that they will improve their material standard of living, the level of fuel carbon emissions which will eventually be seen as reasonable for Australia may be about 20% of the present amount. Thus dealing with global warming will not simply require minor cosmetic changes, but a fundamentally different approach to the use of energy.

There are three broad approaches to the task of reducing the emissions of carbon dioxide. We can reduce the level of energy use by improved efficiency or conservation measures, we can use carbon based fuels which emit less carbon dioxide per unit of energy, and we can make increased use of non-carbon fuels. These are not mutually exclusive; indeed, a sensible strategy to reduce emissions will include elements of all three approaches.

The Australian government set up in 1990 a range of working groups to explore ecologically sustainable development (ESD); the greenhouse effect was one of the key issues for these working groups. The reports of the Ecologically Sustainable Development Working Groups (see ESD Working Groups 1991a) can be seen as a timid first step in the direction of developing a strategy to meet our needs without unacceptable long term effects. The groups concluded that it will be difficult to reduce the rate of burning fuels by 20% by the year 2005. They argued, dubiously on the basis of rather naive economic notions, that to do so might cause considerable economic hardship. This questionable conclusion has been used to argue that the government should back down on its commitment to the 'Toronto 20%' target, although other studies suggest that there would be considerable economic benefits from use of more efficient technologies (Commission for the Future 1991; Lowe 1992). It should also be noted that this approach implies a sovereignty of economic factors above all others, even the capacity of global ecosystems to support the human population.

Improving efficiency

In Chapter 3, Saddler has examined various estimates of the scale of savings which could be achieved by efficiency measures in industry and buildings. While there are disagreements about the exact scale of the savings which are economically viable under current conditions, there is broad agreement that large savings can be achieved by using measures which are cost effective today. Using as the criterion of cost effectiveness a discount rate of 8%, Saddler concludes that carbon dioxide emissions could be cut by over 30% compared with a constant-efficiency future. It should be noted that this calculation assumes no change in the mix of supply technologies and no technical innovations in energy use; it is based entirely on existing technology. On the other hand, Saddler's estimate assumes that the business as usual scenario involves the sort of exponential growth commonly seen by economists as an inevitable consequence of future economic development. A slower rate of increase in energy use will give smaller percentage savings, but lower total emissions. Saddler's chapter shows that a realistic strategy for moving to a sustainable energy system should place a major emphasis on improving the efficiency of end use. This approach is a genuine no-regrets strategy, bringing economic benefits as well as reducing unnecessary emissions of combustion products.

The key policy issue is the identification of strategies which will enable these savings to be achieved. As the ESD Working Groups report on energy use observed (ESD Working Groups 1991b), the current rates of adoption of energy saving technologies will not produce the desired scale of reduction in fuel use or emissions; 'there will need to be well-designed market intervention'. A range of intervention measures could be considered: advisory services, stricter efficiency standards or other regulatory devices, fiscal incentives, easier access to capital for more efficient technology, advertising or education to encourage changes in behaviour. For example, the ESD Working Groups recommended mandatory energy efficiency labelling of all major appliances, the phasing in of minimum standards of efficiency and promotion measures to heighten consumer awareness of the importance of choosing efficient appliances. In similar terms, tighter building codes could make the stock of commercial and residential buildings less wasteful, while measures to compel disclosure of operating costs would establish a market in more efficient space. As these examples show, the policy measures to improve the efficiency of energy use need to be developed with the particular application in mind. The measures recommended by the ESD Working Groups report on energy use were, in the view of that body, 'clearly justified by the economic, environmental and social benefits achieved' (ESD Working Groups 1991b). It also concluded that the steps it envisaged would not harm the competitiveness of Australian industry nor slow the rate of economic development.

A crucial problem in Australia is the level of commitment in the relevant areas of government. As early as 1990, the Commonwealth government committed itself to a package of measures to improve energy efficiency as an interim response to community concern about global warming. As this chapter was being completed, a damning report by the Auditor-General showed that action to achieve the planned savings was well behind schedule (Australian National Audit Office 1993). Even in their own use of energy, the government has failed to set an example to the rest of the community. Improving the efficiency of energy use in manufacturing industry had been neglected. Staff who should have been allocated to the program were still heavily engaged in other work. The overall conclusion was:

> Australia has a poor record of energy saving. Market research and technical studies indicate there is a significant untapped potential to save money and resources and stem carbon dioxide emissions. We are among the world's largest greenhouse gas emitters on a per capita basis. Our cars are among the world's most inefficient in terms of fuel consumption. There is scope to establish benchmarks to ensure our industry is fully aware of competitive opportunities adopted overseas. In terms of construction practices, there was some evidence that we are behind comparable overseas countries in framing requirements to consider energy use. The Department's research showed that few people were aware of the need to save on energy. The public is still unaware and sceptical about Government programs and pronouncements on the topic.

The report noted that there has been little progress on the development of national schemes for energy rating of houses or domestic appliances. Overall, the conclusion was that there seems no strong commitment by government to its stated goal of improving the efficiency of energy use. As similar comments could be made about many State governments, it is tempting to conclude that the problem has structural roots. In most jurisdictions, the department responsible for energy issues has been part of the body responsible for mining and related mineral development. Thus the corporate culture is attuned to the extraction of energy minerals and promotion of their use, rather than to conservation. As an extreme example, the Queensland Department of Minerals and Energy was still saying to the 1991 Fitzgerald Inquiry that sand mining on Fraser Island should be reconsidered, while the entire department was able to spare only one or two people to work on matters related to energy conservation. The trend in government has been to amalgamate small departments into large and unwieldy units, but the report of the Auditor-General shows that energy efficiency is not getting the attention it needs to be effective. It is more likely that progress would be made if each government had a separate Energy Efficiency Agency, with its performance being measured by the reduction in community energy use. There could even be scope for an element of performance pay, or for contracting out these services to private corporations on a payment-by-results basis. It is clear that present structures are failing to deliver the existing potential for savings; even measures which are clearly cost effective are not being implemented. Structural reform appears a key to change.

Transport

As Peter Newman shows in Chapter 4, we should pay attention to the way the layout of buildings constrains the pattern of movement in our cities. The current pattern of urban transport fails on all three criteria of sustainability. It is rapidly using fuel resources in a way which is not sustainable. It is producing pollution levels which are often unacceptable, and it contributes to the breakdown of the social fabric. A sustainable transport system would involve much less use of cars, much more use of public transport, and much more cycling and walking.

In terms of carbon dioxide emissions, road transport accounts for about four times as much as rail, sea and air together, with the vast majority of the emissions from road transport occurring in urban areas (Watson and Watson 1990). Other exhaust gases, such as carbon monoxide and oxides of nitrogen and sulfur, produce serious air pollution problems in our cities (Simpson and Lowe 1991). Gasoline use in our cities is about double the average level for the cities of western Europe and about four times the level for such prosperous Asian cities as Singapore and Tokyo (Newman and Kenworthy 1989; see also Chapter 4). Only the USA has urban areas in which gasoline use per head is greater.

One part of the solution would be to make more use of bicycles for urban transport. There are four sound reasons for doing this. They save fuel, they reduce emissions, they improve community health, and they bring social benefits. Cycling uses one-third the energy of walking, about one twenty-fifth the energy of public transport and about one fiftieth the energy of an average private car (M. Lowe 1989). Each change of a commuter trip from car to bus or cycle makes a contribution, for each saving of a litre of gasoline is equivalent to about three kilograms less carbon dioxide emitted (I. Lowe 1989). The health benefits occur at two levels. Only about 5% of adult Australians exercise sufficiently to maintain basic heart–lung fitness, while a much greater percentage are overweight to an unhealthy extent (Hetzel and McMichael 1987). Increasing the number of adults cycling would increase the fitness of the population. Despite the poor facilities for cycling in Australia, it is still safer than other modes of private transport for the group most at risk of road accidents in Australia, young men. In terms of distance travelled, a cyclist is about half as likely to have a fatal accident as a car driver, about one-third as likely as a pedestrian and less than a tenth the risk of a motor cyclist. Since a significant fraction of the fatal accidents involving cyclists also involve other vehicles, better facilities for cycling would make it an even safer means of transport (Robinson 1991).

The area of greatest potential for energy savings is the pattern of use of passenger transport vehicles. As Newman identified in Chapter 4, Australia has both a high level of reliance on the private car and a low level of car occupancy. This shows there is a need for strategies to reduce car use or increase the average number of occupants per car. While most emphasis usually goes to encouraging public transport, more attention should be given to measures to increase the number of passengers per car. The fuel efficiency of cars per passenger kilometre is only half that of public transport, despite the low occupancy rate, so cars with three occupants are more fuel efficient than the public transport system. It could reasonably be argued that the most significant transport resource in urban areas is the unoccupied seats in private cars. The ESD Working Groups report on transport (ESD Working Groups 1991c) recommended that State and local governments should consider measures to require car pooling. Such measures are common in other nations. Even the USA has urban freeways which are restricted in peak hours to vehicles with three or more occupants. Some cities in Australia have taken the timid first step of denoting one lane as reserved for high occupancy vehicles, but there is not yet even a whole-hearted commitment to this limited approach.

We are a long way from the cultural shift needed to adopt more stringent restrictions on car use. We have yet to take seriously the issue of travel demand management, and existing analytical tools do not give an adequate basis for predicting the effects of specific policy changes (Luk 1992). There is little prospect of change being effected by pricing, however, as a significant fraction of urban commuters are now totally insulated from price signals by their employers. This practice has spread to the public sector, with senior appointments now routinely advertised as having a salary package which includes provision of a vehicle.

A recent review of options concluded that the three factors crucial to improving the fuel efficiency of urban transport are vehicle efficiency, the level of public transport use and the attitudes of urban travellers (Moriarty and Beed 1992). Real change is only likely to occur when driving alone in a car attracts the sort of social opprobrium now directed to smoking in shared space. To put that sort of value shift in perspective, the change in social attitudes toward smoking tobacco has been very rapid in recent years, so it is not fanciful to speculate that similarly dramatic change could occur in the view taken of car use.

The long term level of transport energy use will clearly be strongly dependent on planning decisions, so the whole issue of urban and regional planning has attracted much recent attention. Newman's work has been of great importance in showing the way decisions about urban density and transport infrastructure affect the options available (Newman and Kenworthy 1989; Newman and Kenworthy 1991; Newman et al. 1992). With a growing population and shrinking average household size, urban planning decisions will have a powerful impact on the range of options available for transport. Some transport authorities are still firmly locked into the old paradigm of road construction. As recently as 1989, Queensland still had a Department of Main Roads, naturally seeing its job as constructing more large roads and encouraging their use; while the department has disappeared into a Department of Transport, the mindset is still alive and well. The Brisbane City Council has introduced transit lanes on some major roads under its jurisdiction. But the State department has stolidly resisted such progressive moves on the major arteries it controls, even when road widening schemes provide the opportunity to add a transit lane without reducing the provision for other road users. By contrast, the Department of Road Transport in South Australia has established a working group to look at options for ecologically sustainable development and is taking its long term responsibilities seriously.

There are, Moriarty and Beed noted, no easy solutions to the problem of urban transport emissions. With a growing population and a relatively inflexible infrastructure, options are limited in the short term. Clear priorities should be the promotion of public transport and serious targets for vehicle fuel efficiency improvement. For the longer term, urban and regional planning is the key to keeping transport emissions down to acceptable levels. There is little prospect of achieving the sort of reductions needed without significant changes in public attitudes toward transport options. Milbrath (1989) argues that climate change will be 'the most insistent and persistent teacher', spreading community acceptance of the need for a fundamental change to our pattern of resource use. Without such community acceptance, there is no prospect of changing the pattern of transport energy use to one which could be sustained.

Renewable energy technologies

As Chapters 5–7 show, there are many energy supply technologies which are potentially renewable, being based on harnessing solar energy directly or indirectly. This does not mean that they are environmentally benign, as the technology for converting solar energy into usable forms makes its own resource demands and has its own environmental impacts. That being said, increased use of these technologies would reduce greatly the environmental impacts of energy conversion and use. Renewable energy systems can be used to supply heat, electricity or transport fuels.

The main impediment to the use of renewable forms of electricity is cost. A recent government publication estimated that it would be possible to supply the entire electricity demand of eastern Australia using renewable forms, but at an average supply price in the range $A0.25–0.35 per kilowatt-hour; the average price from the present mix of coal-fired and hydro-electric stations is $A0.085/kWh (Stevens 1992). However, this is an extreme case; the same report envisaged that the supply of 30% of Australia's electricity from solar and wind energy by 2020 would only increase the average price from $A0.085 to 0.095/kWh. Such a minor increase may well be politically acceptable today. Stevens also suggested that it may be possible by 2030 to base electricity supply totally on renewable forms of energy at a cost about 50% more than the present mix; again, that may well be a politically acceptable price for clean energy. These estimates are partly dependent on future developments in the technologies discussed in this volume (Chapters 5–7) by Mills and by Diesendorf, but Stevens argues that the conclusions are robust because several combinations of solar technologies and associated equipment have the potential to achieve similar cost targets. See Table 9.1 below.

Table 9.1 *Present costs and future estimates for various technologies (cents)*

Technology	Supply cost (c/kWh) 1992	2010	2030
Pulverised coal	3.5		
Combined cycle gas	4.5		
Methane from landfill	5.5		
Bagasse	5.6		
Wood	6.3		
Wind	7.1	4.9	4.1
Solar thermal	21.0	6.7	5.2
Photovoltaics	44.0	9.0	5.3
Battery storage	33.0	7.0	7.0

As discussed in earlier chapters, a practical power supply system using renewable energy technologies would require some form of storage. Alternatives to batteries include pumped storage hydro schemes and production of hydrogen as a storage medium. By some measure, the cheapest form of storage is the existing network of hydro-electric stations. While there are certainly environmental objections to more such schemes, there is no reason to refrain from using the existing stock, especially as it facilitates the expansion of other forms of renewable energy.

The second significant impediment to the adoption of renewable energy technologies is the gross overcapacity of most State grid systems. This excess capacity, constructed at huge public cost, resulted from planners using the outmoded model of exponential growth in demand, ignoring warnings of the obvious problems (Lowe 1977). Despite early retirements of expensively refitted power stations and deferment of planned new capacity, the electricity grids of eastern Australia have so much spare capacity that it will be difficult to justify new power stations before the end of the century. The only possible exception is Queensland, where the old assumptions are still being used to argue that the State needs to construct a new hydro-electric scheme in the wet tropics. While the demand case for the Tully–Millstream project is tenuous, the same case may be used to argue for other pilot projects not involving the same capital outlay and environmental objections.

The third major obstacle to use of renewable energy technologies is the dominant attitude of decision makers, most of whom are not comfortable with the new systems. This is attributable to the lack of control over wind or sunlight; the technical challenges of harnessing these forms of energy are not trivial, but no greater than the problems of some other energy technologies. The hostility to renewable energy technologies is manifest in various forms, including opposition to use of solar panels for domestic hot water. This technology has been economic for much of the Australian mainland for more than a decade, but its use has been vigorously opposed by many energy authorities. The very low level of use in Queensland is directly attributable to this sort of opposition; a study ten years ago showed that the economics of solar hot water were better in Queensland than in Western Australia, but solar had 5% of the Queensland market compared with 25% in WA (Lowe et al. 1984). As shown in overseas studies, social attitudes are crucial to the spread of such new technologies (Stern and Aronson 1984). Acceptance by the public is based on social attitudes, but is also affected by such factors as access to capital. That is a serious barrier to the adoption of solar domestic hot water. A high priority should be the introduction of a scheme to allow the solar technology to be paid off over its lifetime from savings in electricity bills. To the contrary, many utilities are still encouraging the use of off-peak electricity for water heating by advertising and, in some cases, direct financial subsidies.

Although electricity utilities are theoretically under the control of State governments, there have been few attempts to direct the utilities to take seriously such modern ideas as least cost planning or demand-side management. Even making judgements about the appropriate fuel for particular purposes is usually left to the utilities, which typically base their final decisions on technical and economic factors. The attention to demand-side management by the State Electricity Commission of Victoria is an exception, attributable to direction from the former Victorian Labor government. Most governments are so committed to primitive notions of deregulation and market forces they are reluctant to intervene in this way.

Two sorts of renewable energy technologies may be considered for use as the energy vectors of transport systems. Alcohol fuels can be considered as proven technology, since they have been used to some extent for decades (Davis 1991). Ethanol and methanol can be produced from plant materials and blended with gasoline or diesel for use as fuel extenders, or used in pure form in modified internal combustion engines. There are problems of scale in seeing either as anything but a fuel supplement. The entire sugar production of Australia would, if fermented to make ethanol, only produce about 10% of the gasoline demand. Growing plant material to produce enough methanol for our transport fuel demand would require a land area comparable to that now used for all agricultural purposes. Either scaling up sugar cane production or creating a vast coppicing industry would pose considerable environmental problems. The most likely use of alcohol in the near future is the possibility of returning to the use of ethanol as an octane enhancer, allowing the removal of tetraethyl lead from gasoline. In 1989, the then government of Queensland indicated it would give preference in its fuel contract to an alcohol blend, but the proposal disappeared with the change of government later that year.

The other possible new technologies for transport would use solar electrical energy, either stored in batteries to power electric vehicles or used to produce hydrogen for use as a transport fuel. Both these technologies are currently more expensive than the internal combustion engine, but both result in much cleaner air. There is pressure to consider them as ways of preventing the air quality problem becoming a serious health risk in urban areas. Either technology would require a massive investment in energy supply infrastructure, as well as needing to overcome the public hesitancy about new systems. In those terms, the electric car has the advantage of being seen at expensive resorts and golf courses, while there is still an unfounded public perception that hydrogen is not safe.

As discussed in this section, there is much that governments can do to promote renewable energy technologies. The most serious impediment is the widespread naive belief in market forces.

Market forces?

The basic premise of 'economic rationalism' is that decisions are best left to the market. Markets are generally recognised as being an economically efficient means of allocating resources. Representing the sum total of consumer choices, they respond to changing demand patterns much more rapidly and flexibly than systems of central control. Despite the superficial appeal of this approach, it has some fundamental shortcomings.

As a basic consideration, many aspects of consumer behaviour are driven by factors other than economics. We all make choices which reflect our emotional needs, our spiritual values or our response to social pressures; it is just naive to assume that all our decisions are those of an economically rational consumer seeking to optimise her or his marginal utility. Secondly, a typical consumer is not in possession of the information which would be needed to make economically rational decisions. Very few people know the rate at which different appliances use energy or the real cost of using a car for short trips. Further, there exists an entire industry – the advertising industry – with the express purpose of trying to persuade consumers *not* to make economically rational choices. There are also some basic limitations on the price mechanism.

If we wanted to leave decisions about energy use to the market but encourage moves toward ecologically sustainable patterns of development, we would have to ensure that the price signals give consumers the right message. Current fuel prices do not include such factors as: allowance for resource depletion; the costs of long term environmental damage; the costs of broad social effects; or the macroeconomic consequences of choices.

Thus, fuel prices do not include any form of compensation to future generations for the depletion of resources, do not include the costs of such effects as acid rain or climate change, do not account for such social effects as the impact on patterns of social life, and do not include any form of provision for the impact of fuel use on such factors as our balance of trade, our self reliance and the strength of our currency. Further, in most of these cases there is no prospect of a convincing algorithm which would allow us to build these factors into the price structure. For that reason the market prices of fuels are not likely to reflect wider considerations, so market choices will also not reflect the wider issues which form part of the consideration of ecologically sustainable development.

Appropriate pricing is one tool that influences consumer behaviour. We can load on to the current fuel prices – which broadly reflect supply costs plus government levies – components to reflect resource depletion, environmental damage or even impacts on natural systems. However, those loadings will be essentially arbitrary, and so will reflect value choices. In the absence of some intellectually acceptable means for deciding the size of such taxes, we could not be sure whether they are even about the right scale. There is unlikely to be consensus about the correct approach, given the arbitrary nature of any loadings and their redistributive effects. Those disadvantaged by a particular choice are very likely to promote other options, making the issue explicitly political. Further, since only the current generation can participate in the market, the wishes of future generations of consumers cannot be reflected in the market choices. Even if prices do not discount the future explicitly, the needs of the future will always be discounted by their inability to be influencing the allocation of resources. Inequalities in market power between different individuals, different groups, different regions and different nations also raise issues of equity.

A final difficulty arises from the understanding that more responsible use of natural resources is not enough to ensure ecologically sustainable development; we must also sustain biodiversity and ecological integrity. It is difficult to see how these issues can be left to a market approach, even in principle; how could we put a fair price on an endangered species or an ecosystem threatened by human activity? Our actions clearly affect other species, but they are unable to participate in any market process to determine priorities. These issues can only be taken into account by thoughtful and unselfish actions; there is no prospect of an optimum outcome resulting from the blind pursuit of self-interest by billions of individual humans. Moving to a system of energy supply and use that will be genuinely sustainable will require a purposeful strategy. It will need a judicious mix of education, financial inducements and regulation. That means we need a more sophisticated approach to policy than we are accustomed to seeing from governments in Australia.

The way ahead: policy options

Several of the technologies discussed in this volume are already economically viable, even by the narrow criteria of 'dry' economics. Others are on the brink of competitiveness. The main barriers to widespread implementation are political and institutional. Some of these barriers were being removed at the time of writing by institutional change. For example, the terms of reference of electricity utilities constituted one barrier to the use of renewable energy and conservation technologies. The moves to 'corporatisation' of these utilities might make the tilt in the playing field less pronounced. The discussion paper canvassing options for the Queensland industry suggested that the electricity industry should be charged with earning a reasonable rate of return on capital, should pay tax and should have no preferred borrowing rights. All these changes are likely to reduce the incidence of such uneconomic practices as extending grid electricity to remote customers at enormous cost, when it would clearly be better on economic grounds to provide stand-alone power supplies using solar technology. The benefits of deferring new capital investment may eventually lead utilities to take demand management seriously, as the State Electricity Commission of Victoria is doing. In the short term, the excess capacity will lead some utilities to promote electricity sales; for example, the renamed Electricity Commission of New South Wales (Pacific Power) has a demand management section, but it is still promoting increased sales under the euphemism of 'load building'. It is not a coincidence that utilities in the USA, operating on a commercial basis, have been much more positive about demand management and renewable energy technologies than any in Australia. These measures make good economic sense; they have been delayed by the institutional arrangements for electricity supply in Australia.

Altering those institutional arrangements may encourage private generation of power with provision for the supply of surplus electricity to grids. Such systems have been impeded by the lack of a realistic buy back rate for surplus power. A cogeneration scheme for some Melbourne hospitals was made economically feasible by an agreement negotiated with the Victorian electricity authority for supply of peak electricity at the avoided cost. This could be seen as a model for further arrangements of this type, although the self-imposed impotence of governments to direct the affairs of utilities makes it less likely that there will be a rush of imitations. The absence of similar agreements in Queensland has been a constant impediment to the supply of grid electricity from the surplus energy available in sugar mills, which burn bagasse as their source of energy. The bagasse is an inevitable by-product of the sugar crushing process; it is also a renewable fuel which puts no net carbon into the atmosphere. On these grounds, its use should be encouraged.

A second barrier arises from the inequitable financial basis for investment in efficiency measures or renewables. Where a utility is content with a payback time of thirty years, householders need a much more rapid return on any investment, as the expected life of solar panels is not much more than ten years. More importantly, the average householder moves about every five years and so any investment with a longer payback time is unlikely to be attractive. Low income groups face a capital barrier; a study of purchasers of solar hot water systems showed a level of total household income below which there were no users, presumably because the capital cost was unobtainable (Lowe *et al.* 1984). As discussed above, this problem could be overcome by financing arrangements. An obvious parallel is the purchase of cars. Very few people would be able to afford a new car if the entire cost was required at the point of sale; most purchasers use a hire purchase scheme by which the cost is spread over the life of the vehicle. Similar arrangements for solar hot water systems would make it available to many more householders.

A related problem is the attitude of industry to investment in new equipment. Particularly during a recession, with cash flow a serious problem for many companies, short payback times are demanded. Thus new technology which would recover its cost in three or four years is unlikely to be seen as an attractive investment in a climate of fiscal uncertainty. Options to circumvent this problem include tax incentives or low interest loans to reduce the perceived risks involved.

Another economic obstacle has been the failure to cost the environmental and resource-consuming costs of fossil fuels. In Chapter 8, Common discussed one possible solution, a carbon tax. There have been two political obstacles to such a levy. One is the political power of the resource industries, which mounted a strong campaign in 1992 against the very idea of attempting to reduce carbon dioxide emissions. While there is no intellectual basis for the claims that such a reduction would do overall economic harm, it is easy to generate a fear campaign against proposed changes, especially when there would clearly be a short term impact on some entrenched interests (Industry Commission 1992; London Economics 1992; Lowe 1992). The second obstacle has been the absence of similar moves in other countries. However, serious moves toward carbon taxes or energy taxes in the USA and the EC are likely to put political pressure on Australian governments to follow suit. Any such changes to the taxation system will reduce the current bias against renewable energy technologies and efficiency measures.

A variety of realistic policies would improve the efficiency of fuel use in transport. In Chapter 4, Newman advocates a system of mandatory standards for new vehicle fuel consumption and fleet efficiency. As shown in that chapter, massive fuel savings could be achieved by such measures. A related option is for vehicle registration charges to be tied to energy efficiency in a simple and transparent way, so that consumers see the benefits of using more efficient cars more directly than through reduced running costs. A further possibility is for road users to pay a higher fraction of the costs they impose on the community, preferably at or near the point of use. Road charging schemes are now technically feasible and being trialled; they are likely to change perceptions of road use dramatically. Many drivers see only the fuel cost of running a car, typically about $A0.07 per kilometre, whereas the real cost is about a factor of ten higher. Seeing that cost would have a real impact on behaviour.

The other element of a sustainable transport strategy should be a real commitment to provide a public transport system that is clean, efficient, safe, reliable and affordable. That will not be a trivial undertaking after decades of systematic neglect.

Government strategies to encourage responsible use of materials should also be considered. The use of recycled rather than virgin materials typically involves a significant energy saving. By the end of 1993, all households in Brisbane had been supplied with special bins for paper, glass, metals and plastics to be recycled. The Brisbane City Council has negotiated markets for the materials to cover the cost of the scheme. There is no reason such a system should not be standard across local government areas, where necessary with the support of State governments to achieve the economies of scale needed to make the scheme viable.

Conclusions

It is now possible to talk seriously about the possibility of government commitment to sustainable development because of the growing mood for change in the community. This needs to be set, however, against the limitations on the capacity of governments in democratic systems to take purposive action. The metaphor Etzioni (1968) developed was of society as 'an ocean liner propelled by an undersized engine'; such a vessel would be partly self-propelled but largely at the mercy of the ocean currents, and there would probably be a constant struggle between different groups of passengers about the best use of the limited capacity to determine where the ship is heading. He argued that such a view of the world is a serious impediment to decisive action (Etzioni 1976):

> Policies that truly express the collective good can only result from a give-and-take among numerous 'partisans' (various interest groups and other active groupings). The measure of a good decision is the extent of agreement on it among all those it affects . . .

As Walker (1992) argues, this approach leads to a fundamental difficulty when applied to environmental issues. There is no prospect, even in principle, of what Etzioni called agreement 'among all those it affects', since those affected clearly include future generations. Also, there are structural problems arising from the nature of the modern industrial state; Galbraith (1967) argues that the nature of the modern state means that politics becomes synonymous with economic management, so that economics becomes the final test of public policy. This focus on economic output as a measure of government achievement leads to the state being, in Galbraith's words, 'an instrument of the industrial system'. The general need for policy not to disrupt the economic system clearly reduces the freedom of the state to take action aimed at alleviating long term environmental problems in general. As Walker puts it:

> These factors limit the freedom states have to attend to ecological or other long-term issues. Even when perceived as problems, they will be relegated to second place when the chips are down. Posterity is a poor second to political survival or economic indicators.

There can be little doubt that responding to the problem of global climate change requires social and political change, rather than technological innovation. Several reports, even by bodies as wedded to the existing order as the Australian Institute of Petroleum (1991), have made the point that 'there are many ways in which governments could take positive steps to reduce greenhouse gas emissions while also contributing to overall economic efficiency'. The task was neatly summarised by the Brundtland Report (World Commission on Environment and Development 1987):

> A safe, environmentally sound and economically viable energy pathway that will sustain human progress into the distant future is clearly imperative. It is also possible. But it will require new dimensions of political will and institutional co-operation to achieve it.

In those terms, the problem of policy change has been cast in extreme terms by those committed to the status quo. It is a common tactic of interest groups to seek to portray their sectional aims as synonymous with the national interest. Given a hypothetical choice between economic prosperity and environmental responsibility, governments are always likely to choose prosperity. The business sectors which could be most affected by strategies to reduce emissions of greenhouse gases have sought to establish that such policies would cause overall damage to the fragile Australian economy. Though there is a significant body of evidence suggesting that the economic benefits of a response strategy would be comparable to the negative effects, those elements of business which would be negatively affected have much more political influence than those which would benefit, and have succeeded in capturing government commitment to that position.

For those reasons, there has been little sign of the political will or institutional cooperation needed to develop a sustainable pattern of energy supply and use. Policy is the art of the possible. It represents decisions taken by those in office, under the influence of values and pressures. For that reason, public policy is by no means permanent, since a variety of different interest groups will always be trying to modify policies to suit their particular interests. This was summed up by Latham (1952):

> What we may call public policy is actually the equilibrium reached in the power struggle at any given moment; it represents a balance which contending factions constantly seek to weight in their favour.

In those terms, we need to be constantly aware of the changing nature of political practicality; what is feasible or desirable is influenced by political perceptions of changing public mood. As this chapter was being finalised, there was a dramatic example of this effect. Though historically hostile to renewable energy, the Queensland government responded to the changing mood of the electorate when it decided to set aside up to $A5 million to research and implement renewable energy systems. The money will come from the operating surplus of the electricity industry. The Minister for Minerals and Energy, Mr McGrady, made the following statement in his press release announcing the initiative:

> It is time to get serious about renewable energy. Solar, wind and geothermal energy technologies have made impressive advances. They are now yielding performance simply unobtainable or prohibitively expensive ten years ago. This is the Sunshine State and these advances mean we now must seriously consider the potential of solar energy to contribute to Queensland's energy demands.

An advisory group has been established to recommend on the use of the fund, with the first priority to investigate alternatives to grid electricity for the Daintree region and the remote western town of Boulia. The end result is likely to be a real, practical demonstration of the capacity of renewable energy technologies to supply the needs of a modern community. The value of such a scheme is inestimable, as successful examples are more likely to change perceptions than any number of feasibility studies by experts or action groups.

It is a graphic and timely reminder that we are constantly choosing our future. Just as the Australia of today is the product of the decisions and actions of previous generations, so the Australia of the next century will be shaped by our decisions and actions, individual and collective, thoughtful and thoughtless, conscious and unconscious. Choices of technology will influence our economic options as well as the shape of our future society. The crucial step is the recognition that the future is not something which just happens or a place to which we are irresistibly propelled by the laws of science, but something we are actively creating. There are clear limitations on our choices in some areas, but in others we have an opportunity to shape our future. If we do not know where we want to go, any path will do; if we do know what sort of future we want, it is possible to work toward it. The only responsible course of action for thoughtful people is one which steers us in a direction of development which we can say, with a clear conscience, we believe to be sustainable.

References

Australian Institute of Petroleum. 1991. *Global warming: alternative policy instruments.* Policy paper 1991/1. Melbourne: AIP.

Australian National Audit Office. 1993. *Efficiency audit: implementation of an interim greenhouse response. Department of Primary Industries and Energy energy management program.* Auditor-General Audit paper 32, 1992–93. Canberra: Australian Government Publishing Service.

Commission for the Future. 1991. *Energy futures.* Carlton, Victoria: CFF.

Davis, J. 1991. *A history of the power alcohol industry in Queensland.* PhD thesis. Nathan, Queensland: Griffith University.

Ecologically Sustainable Development Working Groups. 1991a. *Final. report – executive summaries.* Canberra: Australian Government Publishing Service.

Ecologically Sustainable Development Working Groups. 1991b. *Final report – energy use.* Canberra: Australian Government Publishing Service.

Ecologically Sustainable Development Working Groups. 1991c. *Final report – transport.* Canberra: Australian Government Publishing Service.

Etzioni, A. 1968. *The active society.* New York: Free Press/Macmillan.

Etzioni, A. 1976. *Social problems.* Englewood Cliffs, NJ: Prentice-Hall.

Flavin, C. 1992. Building a bridge to sustainable energy. In: Brown, L. (ed.), *State of the world 1992*, pp. 27–45. London: Earthscan.

Galbraith, J.K. 1967. *The new industrial state.* Harmondsworth, Middlesex: Penguin.

Hetzel, B. and McMichael, A. 1987. *The LS factor.* Ringwood, Victoria: Penguin.

Industry Commission. 1992. *Costs and benefits of reducing greenhouse gas emissions.* IC report 15. 2 volumes. Canberra: Australian Government Publishing Service.

Intergovernmental Panel on Climate Change. 1990. *Climate change.* 3 volumes. Reprinted by the Commonwealth of Australia. Canberra: Australian Government Publishing Service.

International Energy Agency. 1989. *Energy policies and programmes of IEA countries.* Paris: OECD/IEA.

Latham, E. 1952. *The group basis of politics.* Ithaca, NY: Cornell University Press.

London Economics. 1992. *The impact of global warming control policies on Australian industry.* London: London Economics.

Lowe, I. 1977. Energy options for Australia. *Social Alternatives.* 1: 63–69.

Lowe, I. 1986. *Facing the future: energy options for Australia.* Research lecture series. Nathan, Queensland: Griffith University.

Lowe, I. 1989. *Living in the greenhouse.* Newham: Scribe Publications.

Lowe, I. 1992. *Costs and benefits of reducing carbon dioxide emissions.* Canberra: Department of the Arts, Sport, the Environment and Territories.

Lowe, I., Backhouse, D.E. and Sheumack, M. 1984. The experience of solar hot water systems. *Search.* 15: 165–167.

Lowe, M. 1989. *The bicycle: vehicle for a small planet.* Worldwatch paper 90. Washington DC: Worldwatch Institute.

Luk, J. 1992. Models for travel demand management – a review. *Road and Transport Research.* 1(3): 58–73.

Milbrath, L. 1989. *Envisioning a sustainable society.* Albany NY: State University of New York Press.

Moriarty, P. and Beed, C. 1992. Policy options for reducing urban transport greenhouse gas emissions. *Road and Transport Research.* 1(2): 76-87.

Newman, P.W.G. and Kenworthy, J. 1989. *Cities and automobile dependence: an international sourcebook.* Aldershot, Hampshire: Gower.

Newman, P.W.G. and Kenworthy, J. 1991. *Towards a more sustainable Canberra.* Perth: Murdoch University.

Newman, P.W.G., Kenworthy, J. and Robinson, J. 1992. *Winning back the cities.* Sydney: Australian Consumers' Association/Pluto Press.

Robinson, B. 1991. *Bicycle usage and safety in Australia.* Canberra: Bicycle Federation of Australia.

Simpson, R. and Lowe, I. 1991. *Environmental impact review: south-east Queensland passenger study.* Brisbane: Department of Transport.

Stern, P.C. and Aronson, E. eds. 1984. *Energy use: the human dimension.* New York: W.H. Freeman and Co.

Stevens, M. 1992. *Renewable electricity for Australia.* Discussion paper 2. Canberra: Australian Government Publishing Service.

Walker, K.J. (ed.). 1992. *Australian environmental policy.* Sydney: New South Wales University Press.

Watson, H.C. and Watson, C.R. 1990. Near and long term prospects for the reduction in the road transport contribution to greenhouse gases. In: Swaine, D.J. (ed.), *Greenhouse and energy*, pp. 320–330. Melbourne: CSIRO Publications.

World Commission on Environment and Development. 1987. *Our common future.* Oxford: Oxford University Press.

Index

217